CARLAT PUBLISHING

Carlat Psychiatry
Psychiatry Practice Boosters, Fifth Edition

Editor: Jesse Koskey, MD
Associate Editor: Zachary N. Davis, M3

Published by Carlat Publishing, LLC
PO Box 626, Newburyport, MA 01950

Publisher and Editor-in-Chief: Daniel Carlat, MD
Deputy Editor: Talia Puzantian, PharmD, BCPP
Senior Editor: Ilana Fogelson
Associate Editor: Harmony Zambrano

The Carlat CME Institute is accredited by the Accreditation Council for Continuing Medical Education to provide continuing medical education for physicians. Carlat CME Institute maintains responsibility for this program and its content. The American Board of Psychiatry and Neurology has reviewed *Psychiatry Practice Boosters* and has approved this program as a comprehensive Self-Assessment and CME Program, which is mandated by ABMS as a necessary component of maintenance of certification. Carlat CME Institute designates this enduring material educational activity for a maximum of four (4) ABPN Maintenance of Certification credits as part of the 2024 course. Physicians or psychologists should claim credit commensurate only with the extent of their participation in the activity. CME quizzes must be taken online at www. thecarlatreport.com.

Carlat Publishing uses artificial intelligence (AI) tools, such as ChatGPT, in various stages of our content creation process, such as editing articles and creating preliminary drafts and outlines. In all cases, our content is extensively revised during the editorial process by human clinicians and by our board of medical experts to ensure quality and accuracy.

Carlat Publishing books are available at special quantity discounts for bulk purchases as premiums, for fundraising, or for educational use. To order, visit www.thecarlatreport.com or call 866-348-9279.

Print ISBN: 979-8-9893264-7-1
eBook ISBN: 979-8-9893264-8-8

1 2 3 4 5 6 7 8 9 10

Table of Contents

Acknowledgments

THIS BOOK COULD not have happened without the support and love of my wife, Kee. I am also grateful for the indefatigable and incredibly smart Zachary Davis, who contributed insight and accuracy in the midst of med school clerkships. Thank you to Danny Carlat, Chris Aiken, Laurie Martin, Jeff Ives, and Ilana Fogelson, and to my amazing colleagues at the UC Davis Department of Psychiatry. —Jesse

The original research updates adapted for this publication were written by the following authors. All writers have attested that they have no financial relationships with companies related to the material they have written about, or any relevant financial relationships were mitigated before publication.

Deepti Anbarasan, MD

Sarah Azarchi, MD

Jeffrey Cardenas, MD

Nina Chandler, DNP, APRN, PMHNP-BC

Dorothy Chyung, MD

Sy Clark, MD

Erin Conklin, MSN, CNP, PMHNP-BC

Aniruddha Deka, MD

Simon M. Dosovitz, MD

Jason Emejuru, MD

Alex Evans, PharmD, MBA

Peter Farago, MD

Sébastien Hardy, PharmD, BCPS

Earth Hasassri, MD

Neha Jain, MD, FAPA

Thomas Jordan, MD

Rushaniya Khairova, MD, PhD

Kathryn Kieran, MSN, PMHNP-BC

Dominic Le, MD

Timothy J. Legg, PhD, PsyD, MSN, MPA, MSc

Alvin Marquez, DNP

Brian Miller, MD, PhD, MPH

Jeremy Mills, DNP, PMHNP-BC

Gaurav Mishra, MD

Richard Moldawsky, MD

Justin Morales, MD

Sahil Munjal, MD

Eli Neustadter, MD, MSc

Gregory Nikogosyan, DO

Destiny Pegram, MD

Michael Posternak, MD

Xavier Preud'homme, MD

Dee Rapposelli

Vishwani Sahai-Siddiqui, MD

José S. Sanchez-Cruz, MD

Talya Shahal, MD

Susan Siegfreid, MD

Avneet Soin, MD

Maryam Soltani, MD, PhD

Glen Spielmans, PhD

Lara Tang, MD

Briana Tillman, MD

Amy Ton, MD

Kate J. Travis, MD

Marilyn J. Vaché, MD

Alicia Watson, MD

Jaime A.B. Wilson, PhD, ABN, ABPP, MSCP

Introduction

EVERY DAY A NEW BATCH of studies arrives in my inbox, fresh from PubMed. I have the pleasure of sorting through them and then working with Laurie Martin and our wonderful team of writers to turn the most clinically helpful and relevant ones into monthly research updates (RUs). Our goal with this new edition of *Psychiatry Practice Boosters* is to make practicing psychiatry easier and more interesting by distilling the most helpful RUs from the last three years into one volume. This includes not just RUs from *The Carlat Psychiatry Report* (*TCPR*), but those from *The Carlat Addiction Treatment*, *Child*, *Geriatric*, and *Hospital Psychiatry* reports, as well as our new *Psychotherapy Report*.

Clinical psychiatry is not easy. Our patients may linger for weeks or months before something finally helps them feel better. Our best practices can fall far short of what patients deserve. However, knowing that our practices are as up to date and evidence based as possible can help bridge that sometimes very emotional gap between expectation and reality. I hope this collection of RUs shores up your knowledge base, inspires you with new possibilities, and sustains your engagement with our challenging and rewarding field.

For this version, I have updated our "Clinician's Primer on Scientific Research" and included a new section, "Staying on Top of the Evidence With PubMed." In it, I share the practices I've found most helpful as RU editor for keeping up with the literature. There's just so much to read! Of course, the whole point of *TCPR* and its sister newsletters is to do some of that work for you, but for those who have interests and needs that aren't entirely captured by our newsletters, or who like to see for themselves what's out there, I hope this addition will be helpful.

One huge update since the last version of this book is the advent of AI. With the caveat that it might be more quickly out of date than any of this new research, I have also updated the "Clinician's Primer" to include AI prompts and resources that I find helpful in my work.

A Clinician's Primer on Scientific Research

THERE'S A REASON you (probably) don't bring *JAMA Psychiatry* to the beach. Scientific research is dense, jargon-heavy, and not something one reads for entertainment. Keeping up with the literature is essential, though, and sometimes you'll want to examine the data yourself.

The briefs in this book aim to help you stay current and make sense of what matters. But when you dive into a paper on your own, having a strategy helps. Here's a systematic approach, adapted from Jeffrey Barkin's "How to Read a Journal Article" (*The Carlat Psychiatry Report*, February 2007).

GENERAL TIPS

Research papers aren't novels—don't read them front to back. Skim, jump around, and revisit key parts. Spend time on tables and figures; often they tell the story better than the abstract.

Think of papers as arguments. The authors are interpreting data to prove or refute a hypothesis. Financial incentives and cognitive biases can skew those interpretations. So, read with healthy skepticism—like you're preparing to debate the author. As you do, keep the following questions in mind.

1. WHO FUNDED THE STUDY?

Studies funded by pharmaceutical companies are significantly more likely to report favorable outcomes for the sponsor's product (Lundh A et al, *Cochrane Database Syst Rev* 2017;2(2):MR000033). This isn't always because of foul play. Industry-funded trials are often large and well designed, and companies tend to invest in promising compounds. But even high-quality studies can have their outcomes tilted by company-paid scientists. For example, researchers might:

- Underdose the comparator drug
- Choose nonstandard outcome measures
- Shift analyses post hoc to emphasize "positive" findings

Bottom line: Industry funding doesn't automatically invalidate a study's results, but it does mean you need to give them extra scrutiny. Look for disclosures (you'll often find them at the end of the study) and weigh the findings against independent research.

2. WHAT IS BEING STUDIED?

Solid research starts with a clear hypothesis and predefined **primary** and **secondary outcomes.** Declaring these up front prevents data dredging—which is when researchers run countless analyses until something turns up as "statistically significant" by chance.

Say a randomized controlled trial (RCT) is testing antidepressant X for major depressive disorder (MDD). Its researchers set the following outcomes:

- **Primary outcome:** PHQ-9 score change after two months
- **Secondary outcome:** Self-reported sleep hours

If there's no PHQ-9 improvement but sleep increases, that's worth noting. But to truly support antidepressant X for sleep, we'd want a study that declared that aim in advance.

Any post-hoc findings—for example, a finding that antidepressant X improved appetite in women under 30—may be random noise unless confirmed in future studies.

Bottom line: Prioritize findings based on predefined outcomes. Be wary of conclusions drawn from after-the-fact analyses.

3. WHO IS BEING STUDIED?

Study samples often do not capture the real-world complexity of clinical practice. Psychiatric RCTs commonly omit:

- Mild or severe presentations
- Medical and psychiatric comorbidities
- Diverse racial and ethnic groups (Buffenstein I et al, *Sci Rep* 2023;13(1):42)

Also key is how studies handle **dropouts,** which are common in psychiatric trials. Different methods lead to different impressions of a drug's effectiveness. Here are some examples of dropout handling:

- **Last observation carried forward (LOCF):** Keeps a participant's last score before dropout. More conservative, which can make a drug seem less effective.
- **Observed cases (OC):** Includes only those who completed the study. Tends to inflate benefit.
- **Mixed model repeated measures (MMRM):** Uses statistical modeling to account for missing data.

Bottom line: Choose studies with populations that resemble your patients. Pay attention to dropout handling—a study that uses OC may overstate its results.

4. HOW IS THE HYPOTHESIS BEING STUDIED?

Design matters. Here's a quick tour of common study types, ranked by general reliability. PubMed's "article type" filter can sort search results by study design. Reporting checklists for various study types can be downloaded from the EQUATOR Network (www.equator-network.org) to assess study quality.

- **Meta-analyses,** especially of RCTs: These combine results from multiple studies for a broader conclusion. However, meta-analyses are vulnerable to "garbage in, garbage out"—bad input tends to lead to bad output. They are best when they include only high-quality RCTs. Use PRISMA guidelines from EQUATOR to check meta-analyses.
- **Systematic reviews:** Comprehensive summaries of studies, with or without pooled data. Can be excellent resources, especially when RCTs are scarce. "Garbage in, garbage out" also applies here, however—as do PRISMA guidelines.
- **RCTs:** This design helps isolate the effect of the intervention from placebo and expectation bias. Participants are randomly assigned to intervention or control groups. The control group can either receive placebo or standard treatment. "Double-blind" is best—neither participants nor researchers know who gets what. Use CONSORT guidelines to assess RCTs.

- **Observational studies:** These studies involve no randomization. Most can be assessed via STROBE guidelines and include:
  - **Cohort studies:** Follow exposed and unexposed groups over time, either prospectively (forward-looking) or retrospectively (backward-looking).
  - **Case-control studies:** Start with an outcome, then look back to exposures.
  - **Cross-sectional studies:** One-time snapshots of conditions or behaviors.
  - **Case series/case reports:** Descriptions of outcomes in a small group or single patient. Hypothesis-generating but not conclusive. Covered by CARE guidelines.

   Bottom line: Meta-analyses and RCTs provide the most reliable evidence. Use filters in PubMed to prioritize them in your searches and cross-reference with a guideline to assess quality.

5. ARE THE RESULTS STATISTICALLY *AND* CLINICALLY SIGNIFICANT?

Statistical significance means a result is not likely due to chance—but that doesn't mean the result is *clinically* significant. Here are a few key terms to know:

- **p-value <0.05:** The standard threshold for statistical significance, but not necessarily a clinically meaningful change.
- **Confidence interval (CI):** Shows the likely range for the true result. A narrow CI suggests precision.
- **Effect size:** Measures how big the effect is.
  - 0–0.2: small
  - 0.2–0.5: modest
  - 0.5–0.8: moderate
  - 0.8+: strong
- **Number needed to treat (NNT):** How many patients must be treated for one patient to benefit. Lower is better.
- **Number needed to harm (NNH):** How many patients can be treated before one patient is harmed. Higher is better.

   Bottom line: p-values alone aren't enough. Look for effect size and NNT to gauge real-world relevance.

USING ARTIFICIAL INTELLIGENCE (AI) TO ASSESS CLINICAL EVIDENCE

Using commercially available AI platforms to summarize scientific studies is like using it to translate documents from a foreign language. The results can seem good at a glance, but you need to look at the source yourself to judge their validity. ChatGPT, Claude, and Gemini (to name a few) can do many things but can't be trusted to do them without "hallucinations," aka mistakes. (And a mental status exam of a hallucinating AI would always note poor insight.) Even medically focused AI platforms like Doximity GPT and OpenEvidence have disclaimers about inaccuracies and limited clinical utility.

It is tempting to upload papers to an AI platform with the prompt "please summarize this," but those broad prompts leave us vulnerable to hallucinations. Try taking smaller bites out of the apple. AI is most helpful when it explains how it arrived at its conclusions, and teaches us something. It can teach the language of scientific papers or statistics, assess the soundness of research, and connect papers to a larger context.

Here are some prompts that have less chance of invoking a folie à deux. (Full disclosure: AI helped refine this list!)

General approaches:

- "Is it accurate to say that [X] is a correct interpretation?"
- "The authors say [X]. Can you explain that?"
- "Please provide page numbers and exact quotes from the source."

Parsing scientific language:

- "Provide a detailed explanation of any figures or tables in the paper, including how the authors interpret these results."
- "Identify any ambiguous language or areas where the authors' conclusions might overstate their data."

Assessing soundness:

- "Were the statistical methods used appropriate for the study design?"
- "Point out any gaps in the data or methodology that would be important for a clinician to consider."
- "Are there areas of the study that could be prone to misinterpretation if taken out of context?"
- "What assumptions are required to accept the conclusions of this paper?"
- "What alternative explanations for the findings are discussed or could reasonably be considered?"

Connecting to a larger context:

- "Summarize how the authors position their findings within the broader literature."
- "Explain any discrepancies or conflicts with previous research, as discussed by the authors."
- "Are there any findings in this paper that conflict with current guidelines or common practices?"

STAYING ON TOP OF THE EVIDENCE WITH PUBMED

You don't need a wall of journal subscriptions to stay informed. PubMed is the most efficient tool for keeping up with psychiatric research. To use it:

- Set up a free account: pubmed.ncbi.nlm.nih.gov
- Create saved searches using filters and keywords that match your area of interest
- Set up automated email alerts for new studies matching your criteria

Other databases to be familiar with are:

- **APA PsycInfo:** Strong for behavioral/psychotherapy research
- **Embase:** Elsevier's broader search tool with PICO interface (institutional access usually required)
- **Google Scholar:** Allows more natural-language searching than PubMed
- **Scopus:** Includes conference papers and non-journal sources

If you're affiliated with a medical school—even as volunteer faculty—you may have access to pay-walled resources. Your medical or institutional librarian can also be a fantastic resource. They can help with search strategies and full-text access. If you don't have institutional access to articles, email corresponding authors, many of whom will be happy to share PDFs.

Bottom line: Curate your own virtual bookshelf using PubMed. With a few smart filters and alerts, you can stay ahead of the curve.

BIBLIOGRAPHY

Gehlbach SH. *Interpreting the Medical Literature.* 5th ed. New York, NY: McGraw-Hill Education/Medical; 2006.

Ghaemi SN. *A Clinician's Guide to Statistics and Epidemiology in Mental Health.* New York, NY: Cambridge University Press; 2009.

Greenhalgh T. *How to Read a Paper: The Basics of Evidence-Based Medicine.* United Kingdom: Wiley; 2014.

ADDICTION PSYCHIATRY

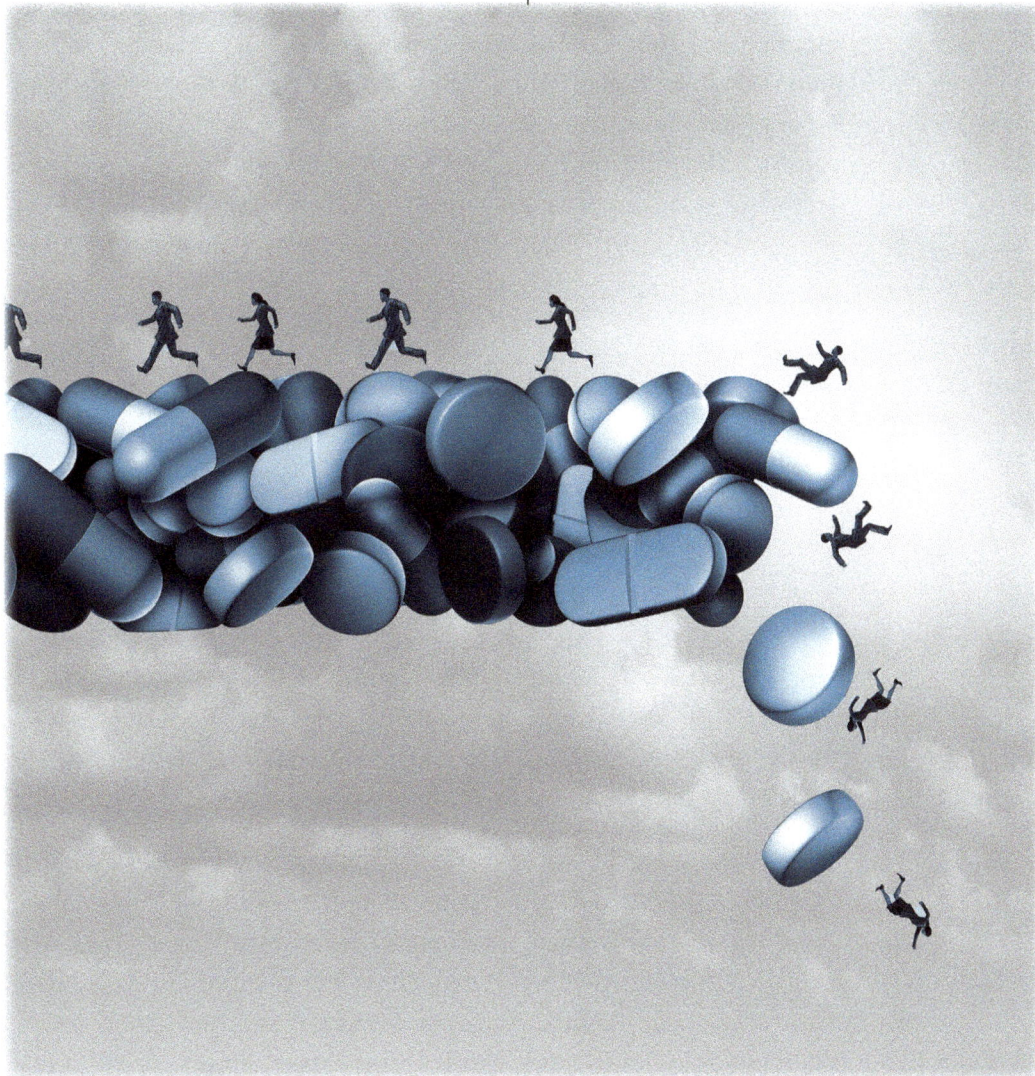

Are MOUD Associated With New Cardiac Arrhythmias?

REVIEW OF: Raji MA et al, *Am J Med* 2022;135(7):864–870.e3

STUDY TYPE: Retrospective cohort

METHADONE, BUPRENORPHINE, and extended-release injectable naltrexone are the three FDA-approved medications for opioid use disorder (MOUD) and are considered gold-standard treatments. Their benefits are well established, including increased treatment retention and significant reductions in both fatal and nonfatal overdoses (Volkow ND et al, *N Engl J Med* 2014;370(22):2063–2066). However, concerns remain about the cardiac safety of these medications, particularly with respect to arrhythmias. Methadone is known to prolong the QT interval, which can increase the risk of torsades de pointes and sudden cardiac death, raising questions about how it compares to other MOUD in real-world settings.

To address this question, researchers conducted a retrospective cohort study of 66,083 individuals with a recent OUD diagnosis but no prior history of arrhythmia. Participants who received MOUD were matched with those who did not, based on demographics, geographic region, and medical comorbidities. The primary outcome was the incidence of new-onset arrhythmia, including QT prolongation, adjusted for baseline characteristics.

Results

Among the study cohort, only 14.1% received MOUD, underscoring the persistent treatment gap in OUD care. Overall, new arrhythmias were slightly more common in the MOUD group than in the non-MOUD group (4.86% vs 3.92%). When comparing the individual medications, however, the findings were surprising. Naltrexone was associated with the highest incidence of new arrhythmia (9.57%), followed by methadone (5.71%), while buprenorphine had the lowest rate (3.81%).

These results appear to conflict with a more recent and larger retrospective cohort study of 144,141 patients with OUD prescribed MOUD, which followed patients for up to 5 years (Wang L et al, *J Clin Psychol* 2023;79(12):2869–2883). That study found methadone carried a significantly higher risk of cardiac arrhythmia—particularly QT prolongation—compared to buprenorphine and naltrexone, which did not differ significantly from one another. Given the conflicting results, further research is needed to clarify these risk profiles. As of 2024, the CDC continues to recommend all three medications as effective treatments for OUD and does not issue specific warnings about arrhythmia risks for any MOUD.

PRACTICE IMPLICATIONS

These findings offer a mixed picture but reinforce a key point: Cardiac arrhythmias are a potential risk when treating OUD with any medication, particularly methadone. While the data on naltrexone are conflicting, buprenorphine appears to carry the lowest cardiac risk and may be a preferred option in patients with preexisting cardiovascular disease or other arrhythmia risk factors. Nonetheless, the benefits of MOUD in reducing overdose and improving retention far outweigh the modest increases in arrhythmia risk seen in these studies. Clinicians should remain vigilant, consider baseline ECGs for high-risk patients, and continue to individualize MOUD selection based on patient comorbidities and preferences.

Buprenorphine Versus Methadone for OUD in Pregnancy

REVIEW OF: Suarez EA et al, *N Engl J Med* 2022;387(22):2033–2044

STUDY TYPE: Retrospective cohort

SINCE THE LANDMARK MOTHER trial, both buprenorphine and methadone have been considered standard treatments for opioid use disorder (OUD) during pregnancy. However, the two medications were not interchangeable—buprenorphine was associated with milder neonatal opioid withdrawal syndrome (NOWS) and shorter hospital stays for newborns, while participants receiving methadone had better treatment retention (Jones HE et al, *N Engl J Med* 2010;363(24):2320–2331). The MOTHER trial was relatively small, with only 175 participants, prompting further investigation using larger datasets. To build on that trial's findings, researchers conducted a large retrospective cohort study using Medicaid data, focusing on neonatal outcomes among infants exposed to buprenorphine versus methadone in utero.

Results

Drawing from a database of over 2.5 million pregnancies, the researchers identified nearly 16,000 pregnant individuals who had been prescribed either buprenorphine or methadone. They found that buprenorphine was associated with significantly improved neonatal outcomes. Compared to infants exposed to methadone, those exposed to buprenorphine had a lower risk of NOWS (52.0% vs 69.2%; relative risk [RR]=0.73; 95% CI [0.73, 0.75]), preterm birth (14.4% vs 24.9%; RR=0.58; 95% CI [0.53, 0.62]), being small for gestational age (12.1% vs 15.3%; RR=0.72; 95% CI [0.66, 0.80]), and low birth weight (8.3% vs 14.9%; RR=0.56; 95% CI [0.50, 0.63]). There were no significant differences between the groups in terms of cesarean delivery or severe maternal complications.

Although the large sample size strengthens the findings, the study's retrospective design comes with important limitations. It lacked information on potential confounders such as medication dose, maternal lifestyle, behavioral health, and social factors. In addition, differences in Medicaid policies across states may have influenced access to care and the characteristics of the study population. Notably, there was no comparison group of patients with untreated OUD, so the relative benefit of medication over no treatment could not be assessed.

PRACTICE IMPLICATIONS

This study provides compelling evidence that buprenorphine is associated with better neonatal outcomes than methadone when used to treat OUD during pregnancy. While methadone may still be appropriate for patients in whom treatment retention is a priority, buprenorphine appears to offer a more favorable risk profile for the infant. Regardless of which agent is chosen, any opioid agonist therapy is strongly preferred over no treatment, as untreated OUD during pregnancy is associated with significantly worse outcomes, including return to use, overdose, withdrawal, and inadequate prenatal care.

Buprenorphine Versus Methadone for Prescription OUD

REVIEW OF: Jutras-Aswad D et al, *Am J Psychiatry* 2022;179(10):726–739

STUDY TYPE: RCT

METHADONE AND BUPRENORPHINE are cornerstone treatments for opioid use disorder (OUD), but most supporting research has been done with individuals using heroin. It remains unclear how well those findings generalize to patients whose opioid use began with prescription medications. To address this gap, researchers conducted the Optimizing Patient-Centered Care (OPTIMA) trial, a pragmatic RCT comparing supervised methadone with flexibly dosed, predominantly take-home buprenorphine for prescription-type OUD (POUD).

A total of 215 participants with POUD were randomized to receive buprenorphine/naloxone (n=107) or methadone (n=108). Methadone was initiated at 30 mg per day, while buprenorphine/naloxone was started between 4/1 mg and 12/3 mg per day. Medication titration and take-home privileges were determined by study clinicians, in alignment with real-world clinical practice. The study was designed as a non-inferiority trial, aiming to assess whether buprenorphine was at least as effective as methadone. The primary outcome was opioid use, measured via urine drug screens collected every 2 weeks over a 24-week period.

Results

The mean maximum dose was 20.3 mg for buprenorphine and 81.8 mg for methadone. Completion rates were slightly higher in the methadone group, with 79 of 108 participants completing the trial compared to 71 of 107 in the buprenorphine group. Medication switching was more common among buprenorphine recipients (22% vs 12%), while take-home dosing was far more frequent in the buprenorphine arm (73% vs 32%). Among those who completed treatment, 24% of urine drug screens in the buprenorphine group were negative for opioids, compared to 18.5% in the methadone group. Adverse events were minimal in both groups, with the most common being mild to moderate withdrawal symptoms in about 6% of participants.

The authors concluded that flexibly dosed, largely take-home buprenorphine was at least as effective as closely supervised methadone in the treatment of POUD. Although methadone had slightly better retention, buprenorphine allowed for more rapid dose titration and early reductions in opioid use. Secondary analyses further supported these findings, showing that buprenorphine was associated with lower cravings—especially early in treatment—and that fentanyl exposure did not significantly affect medication initiation or retention in either group (McAnulty C et al, *Drug Alcohol Depend* 2022;239:109604; Socias ME et al, *Addiction* 2022;117(10):2662–2672).

PRACTICE IMPLICATIONS

This trial supports the use of either buprenorphine or methadone for patients with POUD, confirming that both are effective treatment options. While methadone was associated with slightly better retention, buprenorphine offered similar or better outcomes in terms of opioid use reduction, was better tolerated early in treatment, and provided greater flexibility through take-home dosing. For many patients, especially those who value convenience or face barriers to daily clinic visits, buprenorphine may be the more accessible option. Ultimately, treatment decisions should be individualized, taking into account patient preferences, clinical context, and the feasibility of supervised dosing.

Cannabis Use Frequency and Cannabis Use Disorder

REVIEW OF: Robinson T et al, *Drug Alcohol Depend* 2022;238:109582

STUDY TYPE: Meta-analysis

CANNABIS USE IS COMMON and increasingly legal. What do we know, and how can we advise our patients, about when cannabis use becomes problematic enough to be considered cannabis use disorder (CUD)? Do frequency or potency impact that risk?

To gain more insight regarding frequency, researchers reviewed 6 prospective cohort studies encompassing 40,984 participants between ages 15 and 30, with follow-ups spanning from 3 to 17 years. They identified distinct user groups—nonusers, annual users, monthly users, weekly users, and daily users—and then quantified the rates of CUD across these groups. Unsurprisingly, they found that increased use corresponded with heightened CUD risk.

Results

Specifically, compared to nonusers, relative risk (RR) increased by 2.03 for annual users, 4.12 for monthly users, 8.37 for weekly users, and 16.99 for daily users. Transitioning up one usage category more than doubled CUD risk. To offer a clearer perspective, researchers calculated the absolute risk increase (ARI) as well, which is the overall probability of acquiring CUD. If we assume nonusers had a 0% CUD risk, the ARI was 3.5% for yearly users, 8.0% for monthly users, 16.8% for weekly users, and a staggering 36% for daily users.

This risk profile is further complicated by cannabis potency. A different systematic review published the same year found that high-potency products with >10% tetrahydrocannabinol (THC) significantly increased CUD risk compared to lower-potency products, regardless of frequency of use (Petrilli K et al, *Lancet Psychiatry* 2022;9(9):736–750). Product potency has been steadily increasing over time, with many dispensaries now selling products with THC levels well above 20%.

PRACTICE IMPLICATIONS

We have no formal guidelines with which to quantify patient cannabis use or guide discussions about "safe" use. But asking about daily, weekly, monthly, and annual use is practical and now correlates to a quantified CUD risk. One in three daily users are likely to develop CUD according to the study above. This is twice the risk for weekly users and 10 times the risk for annual users. This evidence also supports counseling patients to decrease not just the frequency of their cannabis use, but the potency of the cannabis as well.

Can Psilocybin Treat Alcohol Use Disorder?

REVIEW OF: Bogenschutz MP et al, *JAMA Psychiatry* 2022;79(10):953–962

STUDY TYPE: RCT

INTEREST IN THE THERAPEUTIC USE of psychedelics has surged over the past decade, with psilocybin—the psychoactive compound in "magic mushrooms"—emerging as a promising candidate for a range of mental health conditions, including depression, anxiety, and end-of-life distress. This study examined its potential role in treating alcohol use disorder (AUD).

In this 32-week, double-blind RCT, 95 adults with AUD were assigned to receive either psilocybin (n=49; dose 25–50 mg/70 kg) or an active placebo, diphenhydramine (n=46; dose 25–100 mg). The study drug was administered in two supervised, eight-hour sessions held at weeks four and eight. In addition, all participants received 12 psychotherapy sessions delivered before, between, and after the dosing sessions.

The primary outcome was the percentage of heavy drinking days (PHDD), assessed through self-reported timeline follow-back interviews. To validate these reports, hair and fingernail samples were analyzed for the alcohol metabolite ethylglucuronide.

Results

Results showed a significantly lower PHDD in the psilocybin group compared to the placebo group (9.7% vs 23.6%; p=0.01). Mean daily alcohol consumption at 32 weeks was also lower in the psilocybin group (1.17 vs 2.26 standard drinks per day; p=0.01). The NNT to prevent a single case of heavy drinking was 4.5—remarkably low compared to the NNT of 12 for naltrexone, a first-line pharmacologic treatment for AUD (Jonas DE et al, *JAMA* 2014;311(18):1889–1900). Psilocybin was generally well tolerated, with only transient elevations in blood pressure and heart rate reported.

Despite these encouraging findings, several limitations must be noted. Outcomes relied heavily on self-reporting, and about half of participants did not complete lab-based confirmations of abstinence. Expectancy effects were likely, as 90% of participants accurately guessed their treatment group. Generalizability may also be limited, as the sample had a median household income of $100,000.

PRACTICE IMPLICATIONS

Psilocybin shows promise as a potential treatment for AUD but remains investigational. Some patients may ask about using psychedelics, including for self-treatment. For those not benefiting from standard care, referral to clinical trials may be appropriate. However, unsupervised use should be discouraged, given the legal risks and the importance of therapeutic support, rigorous screening, and controlled dosing used in research settings.

Daily Alcohol Intake and Risk for All-Cause Mortality

REVIEW OF: Zhao J et al, *JAMA Netw Open* 2023;6(3):e236185

STUDY TYPE: Systematic review and meta-analysis

For years, low-to-moderate alcohol consumption was believed to offer health benefits—particularly for cardiovascular health—but this idea has come under increasing scrutiny. Many earlier studies compared drinkers to abstainers and concluded that ongoing alcohol use was associated with improved health outcomes. However, these "abstainer" groups often included individuals who had quit drinking due to illness, leading to biased comparisons that overstated the benefits of alcohol consumption.

To clarify these associations, researchers conducted a large systematic review and meta-analysis examining the relationship between alcohol use and all-cause mortality. The analysis included data from 107 cohort studies encompassing approximately 4.8 million participants. Participants were categorized into four groups: lifetime abstainers, moderate drinkers (two to three standard drinks per day), high-volume drinkers (three to four drinks per day), and highest-volume drinkers (five or more drinks per day).

Results

The findings challenge the idea that moderate drinking is protective. Compared to lifetime abstainers, high-volume drinkers had a 20% increased risk of death, while the highest-volume drinkers had a 35% increased risk during follow-up periods ranging from 4 to 40 years. No level of alcohol use was associated with reduced mortality risk.

Sex differences also emerged. Two to three drinks per day for women, and three for men, was associated with increased mortality. Across all levels of alcohol use, women had higher mortality risk than men.

The study also highlighted an important methodological limitation: Most of the included studies relied on self-reported alcohol consumption, which is often underestimated. As a result, the actual mortality risks associated with alcohol may be even greater than reported.

PRACTICE IMPLICATIONS

This study represents a significant update in our understanding of alcohol-related health risks. The longstanding belief that moderate drinking might be protective is not supported by current evidence. Clinicians should feel confident telling patients that no amount of alcohol improves longevity.

Deep Brain Stimulation for Severe Alcohol Use Disorder

REVIEW OF: Davidson B et al, *Mol Psychiatry* 2022;27(10):3992–4000

STUDY TYPE: Prospective cohort

DESPITE THE AVAILABILITY of FDA-approved medications, evidence-based psychotherapies, mutual help groups, and various off-label options, alcohol use disorder (AUD) remains a major cause of morbidity and mortality worldwide (Griswold MG et al, *Lancet* 2018;392(10152):1015–1035). For patients with severe or treatment-refractory AUD, novel therapeutic strategies are urgently needed. One such emerging option is deep brain stimulation (DBS)—a neurosurgical intervention, commonly used in Parkinson's disease, in which electrodes are implanted to stimulate targeted brain regions.

In this 12-month observational study, 6 adults (33% female; mean age 49) with severe, refractory AUD underwent DBS electrode implantation targeting the nucleus accumbens (NAc), a brain region critical to the brain's reward circuitry. Participants were monitored over the following year with repeated assessments of alcohol consumption, obsessive-compulsive drinking behavior, anxiety, depression, and liver function. Neuroimaging studies included PET scans to assess glucose metabolism in the NAc and fMRI to evaluate brain activity and connectivity in response to alcohol-related cues.

Results

Over the course of the trial, participants showed consistent improvement in alcohol use and related symptoms. Average daily alcohol consumption declined from 10.4 to 2.7 drinks, though this change did not reach statistical significance ($p>0.05$). Obsessive-Compulsive Drinking Scale scores fell markedly, from 28.7 to 8.3, and anxiety symptoms improved, with Beck Anxiety Inventory scores dropping from 20.3 to 9.3. Liver function also improved, with significant reductions in AST and ALT levels. However, depression scores (measured by the Hamilton Depression Rating Scale) remained largely unchanged. One participant developed an infection at the DBS hardware site, requiring device removal. Although no seizures or hemorrhages occurred during the study, researchers noted these as known risks of DBS. Additionally, interpretation of results may be limited by the lack of a control group, as well as the unusually high motivation likely present in participants willing to undergo brain surgery.

Building on this preliminary work, a 6-month double-blind RCT of NAc DBS for AUD was conducted in which 12 patients with treatment-resistant AUD were randomized to receive either active or sham stimulation. While the primary outcome—continuous abstinence—did not differ significantly between groups, the authors attributed this to limited statistical power rather than a true lack of efficacy (*Transl Psychiatry* 2023;13(1):49).

Further support for DBS comes from a recent systematic review of 26 studies conducted between 2007 and 2023, examining DBS for various substance use disorders. Targeting the NAc was associated with reductions in cravings and substance use across several addictions. However, complete abstinence was inconsistent, and relapse occurred in 73.2% of patients. This result suggests that while DBS may reduce use and improve functioning, it is unlikely to serve as a stand-alone cure (Zammit Dimech D et al, *Transl Psychiatry* 2024;14(1):361).

PRACTICE IMPLICATIONS

DBS represents a promising but highly invasive treatment option for patients with severe, refractory AUD. Given its surgical risks and the mixed—though encouraging—evidence to date, it is unlikely to become a frontline treatment. If eventually approved, DBS will likely be reserved for carefully selected patients and used in combination with conventional therapies.

How Do We Help Depressed Smokers Quit?

REVIEW OF: Cinciripini PM et al, *Depress Anxiety* 2022;39(5):429–440

STUDY TYPE: RCT

Tobacco use disorder and major depressive disorder (MDD) frequently co-occur. This comorbidity is associated with worse health outcomes and reduced success in quitting smoking. Moreover, smoking itself is linked to greater severity of depressive symptoms, creating a difficult clinical feedback loop. To better understand how smoking cessation treatments perform in this population, researchers conducted a secondary analysis of data from the landmark Evaluating Adverse Events in a Global Smoking Cessation Study (EAGLES) trial, focusing on participants with and without depression (Anthenelli RM et al, *Lancet* 2016;387(10037):2507–2520; see the *Carlat Addiction Treatment Report* November/December 2019 for more about the EAGLES trial).

This analysis included 2,635 participants with clinically stable MDD and 4,028 participants without psychiatric comorbidities, the latter referred to as the nonpsychiatric cohort (NPC). All participants smoked at least 10 cigarettes per day. The average age was 47, and 56% were female. Compared to the NPC, the MDD group had longer smoking histories, more prior quit attempts, and higher rates of other psychiatric comorbidities, including anxiety and past suicidal ideation.

Results

Participants were randomized to one of four treatment arms: varenicline, bupropion, nicotine patch, or placebo. All received 12 weeks of treatment and brief smoking cessation counseling. The primary safety outcome was the occurrence of neuropsychiatric adverse events (NAEs), such as mood symptoms, anxiety, or psychosis. The primary efficacy outcome was continuous abstinence during the final 4 weeks of treatment (weeks 9–12), with additional follow-up at 6 months.

Given its antidepressant properties, one might expect bupropion to outperform other medications in the MDD group. However, that was not the case. Varenicline yielded the highest abstinence rates in both cohorts. Bupropion and nicotine replacement therapy were moderately effective, while placebo was least effective. Although abstinence rates were slightly lower in the MDD group compared to the NPC, the difference did not reach statistical significance (p=0.101).

In terms of safety, the overall rate of NAEs was higher in the MDD group than in the NPC (41.1% vs 29.5%). However, rates of NAEs did not differ significantly between medications, and most events were mild. Serious NAEs—those requiring discontinuation or considered clinically significant—were rare, occurring in just 1.8% of the MDD group and 0.7% of the NPC.

PRACTICE IMPLICATIONS

Smoking cessation treatments are both safe and effective for patients with depression. While bupropion may seem like a logical choice due to its antidepressant properties, this analysis suggests that varenicline is the most effective option, regardless of psychiatric status. Pairing varenicline with behavioral counseling remains the most reliable strategy for helping patients, depressed or not, achieve abstinence from tobacco.

Long-Term Patient Outcomes With Buprenorphine for OUD

REVIEW OF: Hasan MM et al, *Am J Drug Alcohol Abuse* 2022;48(4):481–491

STUDY TYPE: Retrospective cohort

OF THE THREE FDA-approved medications for opioid use disorder (OUD), buprenorphine stands out for its proven efficacy, favorable safety profile, and accessibility in office-based settings. Still, one of the most common questions patients ask is, "Do I have to take this medication forever?" It's a fair question—and one for which there have been limited long-term data. To help address this gap, researchers conducted a retrospective, longitudinal cohort study examining how different durations of buprenorphine treatment affect long-term outcomes.

The study included 2,572 individuals with OUD who were prescribed buprenorphine and followed them for up to 3 years after treatment initiation. Participants were divided into 4 groups based on adherence and treatment duration: poor adherence, good adherence for less than 6 months, good adherence for 6–12 months, and good adherence for more than 12 months. The researchers focused on rates of all-cause hospitalizations and emergency room (ER) visits at a pair of key time points: 36 months after starting treatment and 12 months after stopping treatment.

Results

Findings showed a clear dose-response relationship: The longer patients stayed on buprenorphine, the better their outcomes. Compared to those with good adherence for more than 12 months, the odds of hospitalization at 36 months were significantly higher for those adherent for 6–12 months (odds ratio [OR]=1.42; 95% CI [1.09, 1.82]), those adherent for less than 6 months (OR=1.83; 95% CI [1.49, 2.24]), and those with poor adherence (OR=2.71; 95% CI [2.10, 3.51]). The pattern was similar for ER visits: Relative to the >12-month group, odds of an ER visit were higher in those adherent for 6–12 months (OR=1.30; 95% CI [1.01, 1.71]), those adherent for less than 6 months (OR=1.51; 95% CI [1.22, 1.87]), and those with poor adherence (OR=2.71; 95% CI [1.30, 2.19]). These trends held true even when looking at the 12 months following discontinuation.

PRACTICE IMPLICATIONS

At present, there is no clearly defined safe time point for discontinuing buprenorphine or any other medication for OUD. However, this study suggests that longer treatment duration, particularly beyond 12 months, is associated with significantly better outcomes in terms of avoiding hospitalizations and emergency care. When counseling patients, it's worth emphasizing that remaining on buprenorphine is not just about avoiding opioid use—it's also about improving overall long-term health and stability.

Methamphetamine Withdrawal Treatment

REVIEW OF: Wilens TE et al, *J Addict Med* 2024;18(2):180–184

STUDY TYPE: Uncontrolled trial

Between 2012 and 2021, the United States saw a 12-fold increase in overdose deaths related to methamphetamine, amphetamine, and prescription stimulants (Clinical Guideline Committee (CGC) Members et al, *J Addict Med* 2024;18(1S Suppl 1):1–56). Although methamphetamine withdrawal typically does not result in the severe physiologic disturbances seen with alcohol or opioid withdrawal, it is still clinically significant. Patients often experience marked irritability, agitation, depression, and intense cravings—symptoms that can undermine early recovery and may lead to a return to use.

This study evaluated a standardized treatment protocol for methamphetamine withdrawal among psychiatrically hospitalized patients. The approach began with a comprehensive physical exam and administration of high-dose ascorbic acid, based on preclinical evidence suggesting a potential neuroprotective effect against methamphetamine-induced toxicity (Huang YN, *Toxicol Appl Pharmacol* 2012;265(2):241–252). The core of the protocol emphasized behavioral strategies, with medications used as needed and physical interventions minimized. Focus groups were used to assess feasibility and staff perceptions of the intervention.

Staff received training on the effects of methamphetamine, typical withdrawal symptoms, and protocol implementation (see table). The protocol was applied to 23 inpatients—all single men with recent methamphetamine use, the vast majority of whom were experiencing homelessness (91%) and had comorbid opioid use disorder (OUD; 87%). All patients were initiated on medication for OUD (MOUD). The study did not include a control group.

Results

Behavioral interventions alone were sufficient for nearly half of the patients (48%), while the remainder (52%) required medication, most commonly quetiapine. An encouraging 83% of patients completed the protocol, and the average duration of withdrawal symptoms was 2.6 days. Staff reported positive impressions of the protocol, highlighting the value of both behavioral and pharmacologic components, as well as the usefulness of the pre-implementation training.

The study's primary limitations were its small sample size and lack of a control group. While the high adherence rate is promising, it is unclear how much of this success was attributable to the withdrawal protocol itself versus other factors. Notably, all participants received MOUD, which may have influenced outcomes. Given the high prevalence of comorbid opioid use among methamphetamine users, disentangling these effects remains a challenge.

PRACTICE IMPLICATIONS

This small, uncontrolled study offers preliminary support for a structured, hospital-based approach to methamphetamine withdrawal. Although more rigorous research is needed, the protocol provides a practical framework for addressing a withdrawal syndrome that is often overlooked. It also aligns with recommendations from the 2024 *ASAM/AAAP Clinical Practice Guideline on the Management of Stimulant Use Disorder* (CGC Members et al, 2024).

TABLE: Methamphetamine Withdrawal Protocol

Admission Orders	
Comprehensive medical exam	Emphasis on vitals, heart and lungs, dental care, skin excoriation/infection
Ascorbic acid	1000 mg PO BID for 48 hrs
Behavior-Based Orders	
Food and fluid intake	Encourage full meals and hydration; hold meal for later if patient misses meal
Sleep	Allow patient to sleep as they desire, even if it means missing unit activities
Exercise	Encourage physical activity
Medication Orders	
Insomnia	Mirtazapine 15–30 mg QHS
Panic or anxiety	Chlordiazepoxide 25 mg TID PRN; avoid standing benzo administration
Mild anxiety	Diphenhydramine 25 mg QID PRN; hold for disinhibition
Moderate and severe anxiety	Quetiapine 25–50 mg TID PRN
Worsening hallucinations or hallucinations with aggressive or self-harm content	Contact MD/NP

Perinatal MOUD Use and Infant Discharge to Biological Parents

REVIEW OF: Singelton R et al, *J Addict Med* 2022;16(6):e366–e373

STUDY TYPE: Retrospective cohort

FOR PREGNANT INDIVIDUALS with opioid use disorder (OUD), the fear of child welfare involvement can deter them from seeking care. This concern is particularly salient in Native American and Alaska Native communities, where children are disproportionately represented in the child welfare system due to parental substance use. Meanwhile, medications for OUD (MOUD) are known to improve a wide range of maternal and neonatal health outcomes. This study asked a crucial question: Are new mothers who receive MOUD during pregnancy more, or less, likely to be discharged home with their newborns?

Researchers conducted a retrospective chart review across 3 Alaskan hospitals, identifying 193 Native American or Alaska Native patients with OUD who had recently delivered full-term infants. Roughly half of the patients (47%) were receiving MOUD—either buprenorphine or methadone—at the time of delivery. The rest were using non-prescribed opioids, most commonly heroin.

Results

Overall, 70% of infants were discharged to their biological parents. Of the remaining 30%, 13% were placed with a relative, 13% entered foster care directly, and a small number were transferred to another hospital or to a treatment facility with their parent.

Use of MOUD at the time of delivery was strongly associated with infants being discharged to their biological parents (odds ratio [OR]=3.9; 95% CI [1.5, 9.2]; p=0.005). In contrast, prenatal heroin use was associated with a significantly lower likelihood of discharge to a biological parent (OR=0.11; 95% CI [0.04, 0.28]; p=0.0001). Additionally, patients who received adequate prenatal care were more likely to go home with their infants (OR=3.7; 95% CI [1.5, 9.2]; p=0.005) and were more likely to have received MOUD during pregnancy (p<0.001).

PRACTICE IMPLICATIONS

Contrary to a common fear, receiving MOUD during pregnancy was not associated with separation from the newborn—rather, it significantly increased the likelihood of infants being discharged to their biological parents. This study reinforces the importance of encouraging MOUD and prenatal care in pregnant patients with OUD. Promoting treatment engagement may not only improve maternal and infant health outcomes, but also reduce the risk of unnecessary family separation.

Treating OUD in a Multidisciplinary Clinic During the Peripartum Period

REVIEW OF: Mason I et al, *J Addict Med* 2022;16(4):420–424

STUDY TYPE: Retrospective cohort

TREATING OPIOID USE DISORDER (OUD) during pregnancy is critical. Without effective treatment, pregnant patients with OUD face increased risks of overdose, preterm labor, intrauterine growth restriction, and fetal death. Prior research has shown that co-locating obstetric and addiction treatment services improves perinatal outcomes (Meter M et al, *J Addict Med* 2012;6(2):124–130). However, it remains unclear whether these benefits persist into subsequent pregnancies.

To explore this question, researchers conducted a retrospective cohort study of 42 patients with OUD who received care at a multidisciplinary obstetric and addiction clinic during more than 1 pregnancy. These integrated clinics offered prenatal care, initiation and management of methadone or buprenorphine, weekly group therapy, and access to a coordinated team including social workers, psychiatrists, counselors, and nurses. The primary outcome was the use of medications for OUD (MOUD) prior to subsequent pregnancies compared to initial pregnancies. Secondary outcomes included neonatal opioid withdrawal syndrome (NOWS), length of hospital stay, and involvement of child protective services.

Results

The results were encouraging. Patients were more than 6 times as likely to be on MOUD before their subsequent pregnancies than before their first (odds ratio [OR]=6.48; 95% CI [2.52, 16.64]). Improved MOUD adherence was also associated with fewer prenatal urine drug screens positive for illicit substances (64% vs 36%; OR=0.33; 95% CI [0.14, 0.78]). There were no significant differences in secondary outcomes such as rates of NOWS, length of hospitalization, or child protective services involvement.

While the findings are promising, the study's small sample size was a major limitation. In addition, the authors note that NOWS may not be the most informative outcome measure, as it can occur with both MOUD and illicit opioid exposure. Another key limitation was the absence of a comparison group receiving care from separate obstetric and addiction providers. Without this, the specific impact of co-located care cannot be fully isolated.

PRACTICE IMPLICATIONS

Although based on a small cohort, this study suggests that integrated obstetric and OUD care during a first pregnancy may lead to better treatment engagement and reduced illicit substance use in future pregnancies. The findings support the growing call to expand access to multidisciplinary, co-located care for pregnant individuals with OUD.

ADHD

Effectiveness of Unlicensed Stimulant Doses for Adult ADHD

REVIEW OF: Farhat C et al, *JAMA Psychiatry* 2024;81(2):157–166

STUDY TYPE: Network meta-analysis

FINDING THE RIGHT stimulant doses for adult ADHD can be challenging. Practice guidelines—and patients—sometimes suggest exceeding the FDA's approved maximum doses of 60 mg daily for methylphenidate and 40 mg daily for amphetamine-dextroamphetamine. This study aimed to assess the effectiveness and safety of these higher, unlicensed doses.

The researchers included a total of 47 RCTs. 29 tested methylphenidates; 18 tested amphetamines. Collectively they enrolled 7,714 participants with a mean age of 35; 56% were male, and 87% self-identified as White. Researchers calculated standardized mean differences (SMDs) between ADHD symptoms over the course of the studies. They also determined the mean doses associated with 50% and 95% of a maximum change in symptoms (ED50 and ED95) when compared to placebo, and odds ratios (ORs) of discontinuation. They then conducted a network meta-analysis of methylphenidates and amphetamines, comparing placebo, licensed, and unlicensed doses of stimulants.

Results

For methylphenidates, the study showed that while increasing doses generally led to better symptom management, benefits significantly diminished for doses above 40 mg daily (ED50=25 mg, ED95=72.5 mg). The higher doses provided a small benefit above maximal licensed doses (SMD=-0.23; p=0.03) but were associated with decreased tolerability compared to licensed doses (OR=2.02; p=0.01). Likewise, for amphetamines, there was an initial, sharp decrease in symptoms with increased doses, and no improvements beyond 35 mg daily Adderall equivalents (ED50=12.5 mg, ED95=30 mg).

PRACTICE IMPLICATIONS

As with most psychotropics, more is not better with stimulants. This study finds little justification for dosing stimulants above the FDA-approved maximums, particularly with amphetamines, where the benefits maxed out around 35 mg of Adderall (equal to approximately 70 mg Vyvanse) and the risks rose sharply beyond that. For methylphenidate, higher doses bring small gains, but only up to the FDA-approved maximum, and they increasingly come with adverse effects. Patients may report benefits with high doses of stimulants, but the rewarding effects of these medicines make self-reports difficult to interpret. Unless there is clear evidence of functional improvement, consider a slow taper into the therapeutic range.

Reliability of ADHD Screening Tools

REVIEW OF: Harrison AG et al, *J Atten Disord* 2023;27(12):1343–1359

STUDY TYPE: Systematic review

WHEN A PATIENT PRESENTS with a positive score on an ADHD screener, how much weight should we give that score? This study dug into the psychometric properties of commonly used ADHD screening tools to help us interpret their results.

The authors conducted a systematic review, including 20 studies of several self-report measures, such as the Adult ADHD Self-Report Scale. They looked at sensitivities, specificities, positive predictive values (PPVs), and negative predictive values (NPVs) to see how well the tests separated ADHD from both healthy adults and symptomatic but non-ADHD adults. False positive results are expected with screening tools, which are generally designed to be more sensitive than specific in order to catch those who need further evaluation.

Results

Surprisingly, none of the studies employed a comprehensive "gold-standard" evaluation to verify their findings. Many relied on other self-reports, often with high false positive rates, to confirm the diagnosis. PPV scores were notably poor, particularly in clinical settings, where most tests had scores below 30%. In other words, a positive result for most tests had less than a coin flip's chance of correctly identifying ADHD. However, most tests are good at ruling out ADHD. Nearly every measure exhibited an NPV score of 95% or more, signifying that a negative result rarely missed true ADHD, even in the presence of comorbid psychiatric conditions.

Because the DSM criteria for ADHD include domains that are not captured by most self-reports (such as childhood onset and functional impairment), it is unsurprising that diagnosing by self-report alone is not reliable.

PRACTICE IMPLICATIONS
While a negative result on an ADHD screener helps rule out the diagnosis, a positive result doesn't tell us that the patient has ADHD. To diagnose adult ADHD, first rule out other causes. Seek third-party input and look for symptoms that emerged by age 12, are relatively stable, and cause specific problems in multiple areas of life.

ANXIETY AND COMPULSIVE DISORDER

Benzodiazepines, Quetiapine, and Pregabalin for Short-Term Anxiety

REVIEW OF: Munkholm K et al, *Eur Arch Psychiatry Clin Neurosci* 2024;274(3):475–486

STUDY TYPE: Systematic review and meta-analysis

WHEN STARTING A mainstay anxiety treatment like a selective serotonin reuptake inhibitor or therapy, we may also consider adjunctive short-term benzodiazepines or nonbenzodiazepines. The latter, though often off-label, avoid some of the risks associated with benzodiazepines, such as dependency. Do they work as well?

Researchers reviewed RCTs of benzodiazepines, sedating antipsychotics and antidepressants, antihistamines, melatonin, Z-drugs, and pregabalin for treating the first one to four weeks of new-onset acute stress disorder, adjustment disorder, mild to moderate depression, or anxiety. Primary outcomes included the Hamilton Rating Scale for Anxiety (HAM-A), daily functioning, and serious adverse events. The medications were compared via a network meta-analysis.

Results

The search yielded 34 RCTs involving 7,044 patients. Benzodiazepines, pregabalin, and quetiapine all significantly reduced anxiety compared to placebo. Their standardized mean differences on the HAM-A after 1–4 weeks were -0.58 (95% CI [-0.77, -0.40]), -0.58 (95% CI [-0.87, -0.28]), and -0.51 (95% CI [-0.90, -0.13]), respectively, with no significant differences between them.

However, the authors rated the certainty of this evidence as low to very low. Only a handful of trials reported symptom chronicity. Adverse effects were also inconsistently reported, and thus researchers did not draw conclusions regarding tolerability.

PRACTICE IMPLICATIONS

In this study, quetiapine and pregabalin were viable alternatives to benzodiazepines for treating new-onset acute anxiety over one to four weeks. Quetiapine is associated with the risk of arrhythmia, as well as metabolic side effects and tardive dyskinesia. Generally, aim for <150 mg daily, starting as low as 12.5 mg every 8 hours as needed. If choosing pregabalin, consider chronic kidney disease (due to pregabalin's renal excretion) and start at 75 mg twice daily.

Memantine for Trichotillomania and Excoriation Disorder

REVIEW OF: Grant JE et al, *Am J Psychiatry* 2023;180(5):348–356

STUDY TYPE: RCT

PHARMACOLOGIC OPTIONS FOR trichotillomania and excoriation disorder are limited. Selective serotonin reuptake inhibitors, antipsychotics, N-acetylcysteine (NAC), naltrexone, and modafinil have had mixed results. Glutamate plays a role in motor habits, and the glutamatergic modulator NAC improved trichotillomania in a small RCT by Jon Grant and colleagues. In the current study, Grant's team tested memantine, an NMDA receptor antagonist used to treat Alzheimer's disease.

This double-blind trial randomized 100 people with trichotillomania (53%), excoriation disorder (43%), or both (4%) to memantine or placebo for 8 weeks. 86% were women, and their average age was around 31. The 55 participants assigned to memantine took 10 mg/day for a week, then 20 mg/day thereafter. An equal number of participants in each group were in concurrent psychotherapy and/or receiving psychotropic medication. They were excluded only if there had been a change in treatment in the preceding three months. The primary outcome was reduction in a version of the NIMH Trichotillomania Symptom Severity Scale that was modified to include excoriation.

Results

At 8 weeks, 60.5% experienced improvement in the memantine group vs 8.3% in the placebo group (p<0.0001). The NNT for improvement was 1.9. 10.9% in the memantine group and 2.2% in the placebo group had a complete remission of symptoms. There were no serious adverse events reported in either group. Two participants in the memantine arm dropped out due to dizziness.

PRACTICE IMPLICATIONS

These are encouraging results, albeit from a small study. Many people had some degree of benefit, but few totally remitted. We don't have any head-to-head studies of pharmacologic treatments for these disorders, so discuss potential options with patients and consider if any of them cover comorbid conditions (Alzheimer's dementia in this case).

Mirtazapine Augmentation in OCD

REVIEW OF: Mowla A et al, *Int Clin Psychopharmacol* 2023;38(1):4–8

STUDY TYPE: RCT

MIRTAZAPINE IS USED frequently to augment other antidepressants in depression and anxiety, but is this effective for OCD? Mirtazapine's serotonergic actions are complex, leading to speculation that it might even exacerbate OCD.

This 12-week, double-blind, placebo-controlled trial enrolled 61 patients from a single center in Iran. The patients had a primary diagnosis of OCD that did not respond to sertraline monotherapy. Mirtazapine dosing started at 7.5 mg and increased by 7.5 mg weekly until symptoms remitted or side effects limited tolerability. In the end, the mean dosage was 29.56 mg/day. The primary outcome was severity, as measured by the Yale-Brown Obsessive-Compulsive Scale (Y-BOCS). A secondary outcome was response rate, defined as improvement in the Y-BOCS score by 35% or more at week 12.

Results

The mirtazapine and placebo groups started out with similar Y-BOCS scores and mean doses of sertraline (251.37 mg/day and 255.10 mg/day, respectively). Four patients dropped out of each group after enrollment, leaving 22 patients in the mirtazapine group and 23 in the placebo group. Y-BOCS scores improved significantly more in the mirtazapine group relative to placebo at the 4-, 8-, and 12-week checkpoints. At week 12, the average Y-BOCS score in the mirtazapine group was 11.13 +/- 4.27; in the placebo group, it was 18.94 +/- 3.88. Response rates also reached statistical significance at week 12, with 9 responders (40.9%) in the mirtazapine group versus only 1 in the placebo group.

PRACTICE IMPLICATIONS
In this small study, mirtazapine augmentation certainly didn't worsen OCD. Consider it as a second- or third-line option, especially when patients have trouble sleeping.

Which Meds Are Best for Panic Disorder?

REVIEW OF: Chawla N et al, *BMJ* 2022;376:e066084

STUDY TYPE: Systematic review and meta-analysis

THERE ARE MANY effective medications for panic disorder, including tricyclic antidepressants (TCAs), selective serotonin reuptake inhibitors (SSRIs), serotonin-norepinephrine reuptake inhibitors (SNRIs), and monoamine oxidase inhibitors (MAOIs). Researchers used a network meta-analysis to simulate head-to-head comparisons between them. Specifically, the probability that an agent had the best balance of efficacy and tolerability was measured using the surface under the cumulative ranking (SUCRA). The higher the SUCRA, the better the medication's rank.

The investigators searched three databases to identify RCTs that evaluated medications to treat panic disorder with or without agoraphobia. A total of 87 studies involving 12,800 participants and 12 drug classes met the inclusion criteria. Their mean patient age was 35, 64% were female, and the average duration of panic disorder was 6.9 years. The two primary outcomes were remission, defined as no panic attacks for at least one week at the end of a study, and dropouts. Secondary outcomes were adverse effects and symptom scores for depression and anxiety as measured by the Hamilton Depression Rating Scale, Montgomery-Åsberg Depression Rating Scale, and Hamilton Rating Scale for Anxiety.

Results

Compared to placebo, remission rates were significantly higher with benzodiazepines (relative risk [RR]=1.47), TCAs (RR=1.39), SSRIs (RR=1.38), MAOIs (RR=1.30), and SNRIs (RR=1.27). SUCRA rankings echoed these findings, placing benzodiazepines, TCAs, and SSRIs at the top. Buspirone and beta blockers ranked lowest.

Dropout rates were lowest for benzodiazepines and TCA-benzodiazepine combinations, and highest for buspirone and MAOIs. However, adverse events were also most frequent with TCAs and benzodiazepines, and least common with buspirone and SNRIs. Overall, balancing efficacy and tolerability favored SSRIs, especially sertraline and escitalopram. Fluvoxamine, paroxetine, and fluoxetine showed strong efficacy, but also higher rates of side effects. Citalopram had poorer efficacy and greater adverse effects.

Network meta-analyses assume that included studies of varied design are comparable. This analysis was further limited by a high risk of within-study bias, inconsistency and imprecision of findings, and short trial durations.

PRACTICE IMPLICATIONS

Among pharmacotherapy options for panic disorder, this study supports sertraline and escitalopram as the best balance of efficacy and tolerability.

CHILD AND ADOLESCENT PSYCHIATRY

Autistic Drivers Perform Well

REVIEW OF: Curry AE et al, *J Am Acad Child Adolesc Psychiatry* 2021;60(7):913–923

STUDY TYPE: Retrospective cohort

DRIVING IS AN IMPORTANT SKILL for autistic patients; however, clinicians and families worry about the risk of accidents. To address this concern, researchers completed the first longitudinal comparison of autistic and non-autistic drivers. The study linked statewide driver licensing and hospital-reported crash databases in New Jersey, comparing the first 4 years of driving records among 486 reportedly autistic drivers and 70,990 non-autistic drivers; specifically, researchers looked at how many crashes were caused by the driver.

Results

In all, 163 of 486 (33.5%) autistic drivers and 27,018 of 70,990 (38.1%) non-autistic drivers were involved in police-reported crashes. While autistic drivers had similar or lower rates of crashes compared to non-autistic drivers, autistic drivers had far fewer moving violations and were half as likely to crash due to unsafe speeds. Autistic drivers had more accidents from not yielding to the right of way and while making left turns and U-turns.

The authors suggest that these differences reflect challenges with executive skills, visual processing speed, and visual-motor integration. Autistic drivers might be prone to focus ahead and drive where they are looking rather than seeing the whole context, and they may miss road hazards that involve motorists and pedestrians.

The study had some limitations. Autistic drivers were identified by hospital records, not with formal assessments. The authors did not assess the abilities of these drivers (eg, managing the combination of nonverbal communication and visual-motor function needed to anticipate and execute safe left turns). The authors also did not discuss the age at which the drivers obtained their licenses, nor whether they had fewer hours of driving experience. This is important since there is a rapid decline in crash rates after the first few years of driving. Moreover, New Jersey's geography skews toward urban driving, and its driving age requirement (17) is older than most states. This may limit the ability to generalize the study results to younger drivers and those living in rural parts of the country.

PRACTICE IMPLICATIONS

This study suggests that autistic drivers are at least as safe on the road as everyone else. Just as you counsel neurotypical teen drivers about speeding, talk with autistic drivers about how to stay aware of everything happening around them. When appropriate, recommend programs such as Autism Behind the Wheel (https://sellmax.com/driving-with-autism/) that provide specialized support to autistic drivers (Myers RK et al, *Am J Occup Ther* 2021;75(3):7503180110p1–7503180110p11). Several states and the District of Columbia offer optional license designations indicating that license holders have autism or other communication difficulties.

Gender-Affirming Hormones for Adolescents

REVIEW OF: Turban JL et al, *PLoS ONE* 2022;17(1):e0261039

STUDY TYPE: Cross-sectional

APPROXIMATELY 2% OF ADOLESCENTS in the US identify as transgender. Stigma and mistreatment can negatively impact their mental health, and they are at higher risk for depression and suicidality. While many transgender adolescents seek treatment to bring their bodies in line with their gender identities, several state legislatures have banned gender-affirming hormones (GAH) for youths.

Does the initiation of GAH in adolescence improve or worsen mental health outcomes? This study used data from the 2015 Transgender Survey—the largest such survey at that time, with 21,715 respondents—to investigate this important question. The study compared mental health outcomes among individuals who initiated GAH to those of individuals who desired, but never accessed, GAH.

Results

A total of 78% of respondents reported wanting GAH, but only 0.6% and 1.7% reported having access to GAH in early (ages 14–15) and late (ages 16–17) adolescence, respectively. 57% of respondents accessed GAH in adulthood. After adjustments for potential confounders, including age, race, gender assigned at birth, sexual orientation, employment status, and previous pubertal suppression treatment, access to GAH during early or late adolescence was associated with significantly lower likelihood of past-month severe psychological distress ($p<0.0001$) or suicidal ideation ($p=0.0007$) compared with access to GAH in adulthood. Also, access to GAH in adolescence or adulthood was associated with significantly lower risk of suicidal ideation ($p<0.001$) compared to those who wanted but never accessed GAH.

Further studies support and expand upon these findings. A prospective cohort study of 315 transgender and nonbinary youth reported that those receiving GAH showed sustained improvements in depression, anxiety, and positive affect over a 2-year period (Chen D et al, *N Engl J Med* 2023;388(3):252–264). Over a 12-month period, another study found that gender-affirming medical interventions were associated with 60% lower odds of depression and 73% lower odds of suicidality among 104 transgender and nonbinary youths (Tordoff DM et al, *JAMA Netw Open* 2022;5(2):e220978). And a 2023 systematic review of 46 studies confirmed that GAH consistently reduces depressive symptoms and psychological distress in transgender people (Doyle DM et al, *Nat Hum Behav* 2023;7(8):1320–1331).

PRACTICE IMPLICATIONS

GAH has been politicized in some communities. However, the evidence here is clear. Our transgender and nonbinary patients are likely to do better when they have access to GAH, including before adulthood. The results of this study support the Endocrine Society's recommendation that transgender adolescents have access to GAH.

Is It Worth Adding Coenzyme Q10 to Atomoxetine for ADHD?

REVIEW OF: Gamal F et al, *CNS Neurol Disord Drug Targets* 2022;21(8):717–723

STUDY TYPE: RCT

MANY SUPPLEMENTS HAVE BEEN TOUTED to help ADHD. The pathophysiology of this disorder may be associated with oxidative stress, so it is reasonable to consider an antioxidant intervention (Joseph N et al, *J Atten Disord* 2013;19(11):915–924). In this study, researchers tested whether augmentation of atomoxetine with the antioxidant coenzyme Q10 (CoQ10) would further improve ADHD symptoms.

Researchers recruited 60 children aged 6–16 who continued to have ADHD symptoms despite taking atomoxetine for 6 months. Half received placebo and the other half received CoQ10 (1–3 mg/kg/day). Serum levels were not checked in this study; however, the dosage came from a pediatric migraine study that showed increased CoQ10 serum levels using this dosage range from a mean of 0.6 µg/mL to 1.2 µg/mL (Hershey AD et al, *Headache* 2007;47(1):73–80). ADHD symptoms were measured with the Conners Parent Rating Scale-48 before and after 1, 3, and 6 months of treatment.

Results

The addition of CoQ10 yielded a 16% greater improvement than placebo in ADHD hyperactivity, impulsivity, and learning problems after 6 months of treatment. Adverse effects included nausea; however, there was no statistically significant difference in adverse effects between the groups.

Since this study was conducted in Egypt, where stimulants are not widely available, the patient population may be different than in the US, where many people are prescribed atomoxetine if they are unable to tolerate stimulants.

PRACTICE IMPLICATIONS

There is not yet enough evidence to routinely recommend CoQ10 for ADHD. However, many patients with ADHD do not tolerate stimulants, or get only partial benefit from atomoxetine or alpha agonists. For them, we can add CoQ10 to our list of understudied—though inexpensive, likely harmless, and possibly helpful—adjunctives.

Olanzapine and Samidorphan:
A Promising Combo

REVIEW OF: Kahn RS et al, *J Clin Psychiatry* **2023;84(3):22m14674**

STUDY TYPE: RCT

OLANZAPINE IS APPROVED for teens aged 13–17 with schizophrenia or acute mania, and, in combination with fluoxetine, for bipolar depression in children aged 10 and older. It's efficacious but can cause significant weight gain. Olanzapine and samidorphan (OLZ/SAM, brand name Lybalvi) was approved in 2021 for adults with schizophrenia, mania, or mixed episodes of bipolar I disorder (BD). Samidorphan is an opioid antagonist included with olanzapine to reduce weight gain. Is OLZ/SAM effective, albeit off-label, in teens?

In this industry-funded RCT, researchers looked at 428 young adults between ages 16 and 39. All study participants presented within four years of symptom onset of schizophrenia, schizophreniform disorder, or BD. The study assessed the comparative weight-sparing effect of OLZ/SAM versus olanzapine alone. None of the patients were obese at the outset of the study.

Results

At the end of 12 weeks, patients taking OLZ/SAM were less likely to have gained weight than those on olanzapine alone (4.9% compared to 6.8%). Those on OLZ/SAM who did gain weight gained an average of 7.5 lbs (3.4 kg), while those on olanzapine gained an average of 10.3 lbs (4.7 kg). The weight gain was apparent at six weeks and continued until the end of the study. Patients under 30, unlike older patients, experienced a significant weight-protective effect from the combination pill versus olanzapine alone.

Equal numbers of patients experienced mild to moderate side effects, which included sleepiness, dry mouth, and constipation. Both groups had similarly modest changes in metabolic laboratory parameters, including lipids and glucose. More study participants experienced severe adverse effects on olanzapine alone (5.1% compared to 0.9%), but notably, a single participant in the OLZ/SAM group suffered a seizure that was attributed to the medication.

PRACTICE IMPLICATIONS

This is an industry-funded study, and therefore, we are cautious about the veracity of its content. Moreover, the researchers only included children aged 16 and older, and only broke out weight-related data for a subgroup of patients under 30. Nevertheless, that group did benefit from the OLZ/SAM combo. The combination would be a reasonable approach for older teens who have weight-related comorbidities and haven't benefited from an FDA-approved medication. Keep in mind that one patient had a seizure attributed to this medication.

Physical Activity for Depression in Youth: A Closer Look at the Data

REVIEW OF: Recchia F et al, *JAMA Pediatrics* 2023;177(2):132–140

STUDY TYPE: Systematic review and meta-analysis

WE OFTEN LOOK for ways beyond medications and therapy to help manage depression in young people, especially when there are side effects or a lack of engagement. Physical activity could step in as an adjunct, but the 2023 AACAP Clinical Practice Guidelines for children and adolescents with depressive disorders concluded there was "insufficient rigorous evidence of benefit for . . . exercise" (Walter HJ et al, *J Am Acad Child Adolesc Psychiatr* 2023;62(5):479–502). Let's jump into some recent findings on this.

This comprehensive review brought together data from 21 studies, including 17 RCTs. These studies investigated the effects of aerobic-type physical activity lasting at least 4 weeks, though the average was about 22 weeks. The sessions typically lasted 50 minutes and occurred 3 times a week. Researchers were particularly focused on how these activities impacted depressive symptoms in children and adolescents as measured by validated scales.

Results

Overall, 2,441 participants (53% girls, average age 14) took part. Physical activity interventions reduced depressive symptoms to a statistically significant degree, although the mean effect size was small at -0.29 (p=0.004). The NNT was 6. Interestingly, the programs with the most benefit were less supervised, were shorter than 12 weeks, and had participants who were over 13 or were clinically diagnosed.

Another large meta-analysis, published later the same year, also confirmed that physical activity had significant mental health benefits for depressed pediatric patients (Li J et al, *BMC Public Health* 2023;23(1):1918).

> **PRACTICE IMPLICATIONS**
> The evidence for exercise in pediatric depression may not meet the rigorous standards for clinical practice guidelines, but it does suggest that children and adolescents are likely to benefit. The benefits of regular exercise go well beyond pediatric depression. Especially when medications and therapy aren't enough, it makes a lot of sense to recommend that patients make exercise a habit.

Stimulant Treatment Effect on Anxiety in ADHD

REVIEW OF: Soul O et al, *J Child Adolesc Psychopharmacol* 2021;31(9):639–644

STUDY TYPE: Prospective cohort

ANXIETY IS ONE OF THE MANY side effects commonly listed for stimulant medications. Although a 2015 meta-analysis showed that stimulants did not worsen anxiety symptoms in children with ADHD compared to placebo, none of the studies included in that paper used prospective methodology (Coughlin CG et al, *J Child Adolesc Psychopharmacol* 2015;25(8):611–617). Soul and colleagues published the first prospective study examining whether stimulants exacerbate anxiety.

This 12-week study recruited children ages 6–15 who were starting a stimulant for ADHD. Sixty-eight percent of these were stimulant-naive, and the rest were starting a new stimulant. Parents filled out baseline ADHD Rating Scale (ADHD-RS) and Screen for Child Anxiety Related Disorders (SCARED) scales. Eighteen had diagnosed anxiety disorders: 8 separation, 2 generalized, 4 social, 1 panic disorder, and 3 other specified anxiety disorders. The stimulants prescribed included long-acting methylphenidate for 51.8% of participants, with other prescriptions split among osmotic-release oral system methylphenidate, lisdexamfetamine, mixed amphetamine salts, dexmethylphenidate, and immediate-release methylphenidate.

Results

As expected, ADHD symptoms improved over the next 12 weeks, regardless of anxiety disorders. However, global SCARED scores also decreased significantly, from 21.07 +/- 10.51 at baseline to 17.02 +/- 9.70 at 12 weeks. The effect size was 0.12 ($p=0.001$). There were significant decreases on subdomains of generalized anxiety disorder, separation anxiety disorder, and a school-avoidant behavior subscale; however, the effect sizes were similarly small, with none beyond 0.14. About 14% of the patients reported side effects of irritability, tension, and anxiety symptoms, and 12% reported depressive symptoms.

PRACTICE IMPLICATIONS

This study was small, was open-label, and lacked controls. Nevertheless, the prospective, naturalistic design and the use of validated scales provides reassurance that children with comorbid ADHD and anxiety are, on the whole, not likely to suffer from increased anxiety when prescribed a stimulant.

Testing Neurofeedback for ADHD

REVIEW OF: Neurofeedback Collaborative Group, *J Am Acad Child Adolesc Psychiatry* 2021;60(7):841–855

STUDY TYPE: RCT

NEUROFEEDBACK (NF) IS MARKETED DIRECTLY to patients as a "natural" (and expensive) treatment for ADHD. Does it work? A meta-analysis of small studies demonstrated gains for inattention and hyperactivity-impulsivity that persisted at six-month follow-up (Van Doren J et al, *Eur Child Adolesc Psychiatry* 2019;28(3):293–305). An RCT found that NF was more effective than computerized attention skills training; however, it did not use a standard protocol (Gevensleben H et al, *J Child Psychol Psychiatry* 2009;50(7):780–789). The authors of the present study wanted to give NF a more thorough evaluation.

In this randomized, double-blind, placebo-controlled trial, researchers assigned 84 children with ADHD to NF and 58 to sham NF. Subjects were ages 7–10, and 111 were male. They were allowed to be on stimulants, but no other psychiatric medications. The treatment group received an NF protocol for ADHD known as theta-beta ratio (TBR). Both groups had a target of 38 treatments and received counseling on sleep and nutrition. The primary outcome was parent and teacher Conners scores.

Results

Both groups had significant improvements from baseline, but there were no statistically significant differences between the active and sham groups at the end of the protocol, or at 13-month follow-up. The authors speculated that both groups may have benefited from consistent attention to diet and sleep, the EMG biofeedback component that both received, or the reinforcement for an activity requiring attention.

These speculations were validated by the same group's 25-month follow-up. In that report, there were also no significant differences between the active NF and sham groups. The improvements seen in both groups remained stable, suggesting that the benefits were due to nonspecific effects of the treatment package rather than the specific EEG training itself. Interestingly, they noted that the magnitude of improvement was comparable to that seen in the Multimodal Treatment Study of ADHD, further suggesting a significant psychotherapeutic/behavioral effect from the overall treatment approach (Neurofeedback Collaborative Group, *J Am Acad Child Adolesc Psychiatry* 2023;62(4):435–446).

PRACTICE IMPLICATIONS

This is the first large-scale RCT of NF for ADHD. The negative results from both the initial and follow-up studies may be disappointing, but that lack of significant difference between TBR NF and sham is helpful evidence to discuss with families who ask about the treatment.

Ziprasidone for Bipolar Mania in Children and Teens

REVIEW OF: Findling RL et al, *J Child Adolesc Psychopharmacol* 2022;32(3):143–152

STUDY TYPE: RCT

FDA-APPROVED MEDICATIONS for bipolar I disorder (BD) mania in children include aripiprazole, asenapine, lithium, olanzapine, quetiapine, and risperidone. Ziprasidone is a dopamine and 5-HT2A antagonist with a relatively neutral metabolic profile than some other second-generation antipsychotics. It has an FDA approval for BD in adults, but in 2009, Pfizer was denied FDA approval for ziprasidone for pediatric BD. This was due to a lack of long-term data, high QT prolongation, higher rates of adverse effects in children, and high loss to follow-up (https://b.link/2jj8b7). Pfizer was fined for illegally promoting ziprasidone for off-label uses, including for pediatric patients; in the following year, the FDA reported concerns about Pfizer's clinical trials of ziprasidone (https://b.link/x1k5ju; https://b.link/sgkasy).

Then, in 2013, Pfizer funded a 4-week, double-blind RCT for manic and mixed BD episodes in subjects aged 10–17 (Findling RL et al, *J Child Adolesc Psychopharmacol* 2013;23(8):545). In that study, ziprasidone was effective, with a fairly impressive 0.5 effect size; only 1 subject had a QTc interval over 460 msec. In a 26-week open-label extension of that trial, ziprasidone was well tolerated—there were no clinically significant changes in movement disorder scales, BMI z-scores, liver enzymes, or fasting lipids and glucose. This study was a Pfizer-funded replication of that trial. 86 subjects were randomized to ziprasidone, and 85 to placebo.

Results

Ziprasidone outperformed placebo with an effect size of 0.58. This is somewhat lower than other trials of second-generation antipsychotics, which show a pooled effect size of 0.65 (Correll CU et al, *Bipolar Disord* 2010;12:116). There was no clinically significant effect on BMI or metabolic parameters. The main side effects of ziprasidone in descending order included somnolence, fatigue, nausea, extrapyramidal symptoms, and loss of appetite. The discontinuation rates were 26.7% in the treatment group and 11.8% in the placebo group. The mean QTc prolongation for those treated with ziprasidone was a relatively inconsequential 5.44 msec, with only 1 patient moving into the abnormal range (eg, >460 msec).

PRACTICE IMPLICATIONS

Although we are wary of the reliability of industry-funded studies, in this replication study, ziprasidone again showed a reasonable effect size for pediatric mania and is associated with minimal metabolic impact. It is still off-label for pediatric patients, but may be a reasonable choice if FDA-approved options are exhausted or contraindicated.

GERIATRIC PSYCHIATRY

Assessment of Stimulant Use and Cardiovascular Event Risk Among Older Adults

REVIEW OF: Tadrous M et al, *JAMA Netw Open* 2021;4(10):e2130795

STUDY TYPE: Retrospective cohort

STIMULANT USE AMONG OLDER ADULTS has increased in recent years, not only for the treatment of ADHD, but also for off-label indications. We know that stimulants are associated with cardiovascular (CV) safety concerns. So how worried should we be when prescribing them to older people?

In this population-based cohort study, researchers explored the association between a stimulant prescription and the risk of adverse CV events among adults ages 66 and older in Ontario, Canada between 2002 and 2016. The researchers used health care databases to compare 6,457 older adults who received a new stimulant prescription to a control group of 24,853 older adults who did not take stimulants. The primary outcome was any CV event, which the investigators defined as an emergency department visit or hospitalization for myocardial infarction, stroke, transient ischemic attack (TIA), or ventricular arrhythmia.

Results

Those taking stimulants were found to have a 40% increased risk of CV events and a 240% increase in all-cause mortality at 30 days, compared to those who had not received stimulants. Neither was present at 180 days, and by 365 days all-cause mortality risk was actually 30% lower for those taking stimulants. A secondary analysis of individual adverse events found that stimulant initiation was associated with a 300% increased risk of ventricular arrhythmias and a 60% increased risk of stroke or TIA at 30 days. These associations did not persist at 365 days, although at 180 days the risk of arrhythmias was still 300% higher.

The authors noted a possible selection bias in which patients who continued stimulant prescriptions over the long term represented a subset of patients at lower CV risk. Other limitations included missing clinical variables such as smoking history, alcohol use, and BMI, which can affect CV risk. A meta-analysis published the following year evaluated CV risks of stimulants and other ADHD medications for almost 4 million people of all ages. There was not a significantly elevated risk for older adults, although the authors noted that their analysis was comprised of heterogenous studies, and the CIs of those studies were relatively wide (Zhang L et al, *JAMA Netw Open* 2022;5(11):e2243597).

PRACTICE IMPLICATIONS

This study suggests that in older adults, stimulant prescriptions are associated with a significantly increased CV risk. The risk is greatest for the first 30 days. However, these findings should be interpreted with caution, as clinicians may have discontinued stimulants if patients exhibited CV symptoms within that 30-day period. We should review a patient's cardiac history, obtain a baseline ECG, and discuss the elevated risk of stroke/TIA and ventricular arrhythmia with older patients before starting stimulants. Consider closer monitoring during the first 30 days.

Can Physical Activity Offset Cognitive Decline?

REVIEW OF: Desai P, *JAMA Netw Open* 2021;4(8):e2120398

STUDY TYPE: Prospective cohort

DOES EXERCISE PREVENT—or at least slow down—dementia? It's a tantalizing but still unanswered question. Many studies have shown a correlation between physical activity and better cognitive functioning, but not a causative link. Higher cognitive capacity could just as easily encourage people to be more physically active, or cognitive impairment could result in a reduced ability to exercise.

The latest study to weigh in on this issue still doesn't help us with the causality question—but it does measure biomarkers in addition to mental status, which adds "harder" data to the exercise-cognition connection. This study measured blood levels of tau protein, as high levels of tau have been associated with cognitive decline and the progression from mild cognitive impairment to Alzheimer's disease (AD).

Researchers used data from 1,159 older adults who participated in the Chicago Health and Aging Project between 1993 and 2012. Participants were older than 65 without AD at study entry. Participants were included if they had a baseline blood sample measuring total serum tau concentrations and at least two cognitive assessments. The average age of participants was 77, and they were predominantly female (63%) and Black (60%) with a mean educational level of 12.6 years. Participants were divided into 3 groups by self-reported duration of physical activity: little (no exercise), medium (<150 minutes/week), and high (>150 minutes/week). Cognitive function was measured by the East Boston Memory Test, the Symbol Digit Modalities Test, and the Mini-Mental State Examination. Results were adjusted for demographic factors, including baseline APOE4 status and chronic medical conditions.

Results

Participants with high total tau at baseline had a slower rate of cognitive decline if they reported high or medium physical activity compared to those who reported no exercise. Among participants with low tau, physical activity was also associated with slower cognitive decline, but the benefit was smaller.

Limitations of this study included generalizability, as the study only enrolled Black and White participants. Physical activity levels lacked details as to type and intensity, and self-report can introduce bias into the data.

Subsequent research has continued to expand our understanding of the relationship between physical activity and cognitive health. A 2024 meta-analysis of 104 studies with 340,000 participants found a consistent association between physical activity and slower cognitive decline (Iso-Markku P et al, *JAMA Netw Open* 2024;7(2)). However, the protective effect was modest. A different 2024 trial investigated whether 6 months of exercise could alter biomarkers of AD, including phosphorylated tau (p-tau181), in 99 cognitively normal older adults. Despite cognitive improvements in the exercise group, particularly in episodic memory, no significant changes were observed in p-tau181 or in other biomarkers (Sewell KR et al, *GeroScience* 2024;46:5911–5923).

PRACTICE IMPLICATIONS

The relationship between tau, cognition, and exercise is not a simple one. However, this research does confirm the benefits of exercise for cognition in older age. It's good practice to encourage moderate physical activity in all older patients who can tolerate it—especially those with a family history of dementia, a history of depression, or other risk factors for poorer cognition later in life.

Do Psychosocial Interventions Improve Quality of Life in Advanced Dementia?

REVIEW OF: Hui EK et al, *Int J Geriatr Psychiatry* 2021;36(9):1313–1329

STUDY TYPE: Systematic review

THERE ARE SEVERAL PSYCHOSOCIAL TREATMENTS that may improve quality of life (QoL) in patients with advanced dementia. A recent review analyzed the evidence for them.

The authors reviewed RCTs published from 2000 to 2020 studying psychosocial interventions, including physical, cognitive, or social activities, that aimed to improve functioning and well-being in people diagnosed with moderate to severe dementia (a Mini-Mental State Examination score of 20 or less; the most common diagnoses were Alzheimer's disease, vascular dementia, or mixed dementia).

A total of 14 studies enrolling 1,161 adults were analyzed. The interventions included six multisensory stimulation programs, including aromatherapy (a multisensory intervention involving essential oils, such as lemon balm, provided by diffusion or massage); five multicomponent programs (interventions with more than one type of program, such as exercise and music); two exercise programs; and one reminiscence therapy program. (Reminiscence therapy involves recalling pleasurable or meaningful past events via tangible prompts.) The median duration of the interventions was 12 weeks.

Results

Only aromatherapy and reminiscence therapy were associated with significant improvements in QoL ratings (p=0.01 and <0.01 respectively). One study of robotic pets also showed significant improvements in QoL on subgroup analysis of people with moderate to severe dementia (p=0.01), although the subgroup was too small to be convincing. Both the aerobic exercise programs and one of the multicomponent programs (involving aerobic exercise, memory games, and music therapy) had significant improvements in cognition scores (p=0.01, <0.001, and <0.05), but all were either insufficiently powered or had low-quality designs. The aerobic exercise was low-intensity, like cycling 15 minutes daily or walking 30 minutes daily.

Subsequent evidence has further strengthened the cases for both aromatherapy and reminiscence therapy in dementia. A 2024 meta-analysis of 15 RCTs involving 821 patients found that aromatherapy significantly reduced behavioral and psychological symptoms of dementia after a month of treatment (Wang PH et al, *J Am Med Dir Assoc* 2024;25(11):105199). And a 2022 RCT of reminiscence therapy for 148 people with Alzheimer's disease or vascular dementia found significant improvements in quality of life (among other benefits) over a 13-week intervention period (Pérez-Sáez E et al, *Clin Neuropsychol* 2022;36(7):1975–1996).

PRACTICE IMPLICATIONS

Aromatherapy and reminiscence therapy can improve QoL in patients with moderate or severe dementia. Aromatherapy is relatively inexpensive, and while reminiscence therapy requires a trained facilitator, it can be done in groups. Family members might also make use of the core ideas of reminiscence therapy by providing their loved ones with personal photos and keepsakes from home.

Enhancing Antidepressant Efficacy in Older Adults

REVIEW OF: Srifuengfung M et al, *Ther Adv Psychopharmacol* 2023;13:1–14

STUDY TYPE: Review

TREATING DEPRESSION IN OLDER ADULTS isn't just about choosing the right antidepressant—it's also about getting the dose and duration right, and avoiding harmful drug interactions. This population often deals with multiple health issues, cognitive decline, and the risks of polypharmacy, which complicates their treatment. Many older patients end up on antidepressant doses that are too low, or don't continue their treatment long enough to see full benefits. This review highlights strategies for optimizing antidepressant use while minimizing the risks associated with inappropriate medications like anticholinergics and benzodiazepines.

Findings were consolidated from a literature search, prioritizing studies published within the past 10 years, including RCTs, meta-analyses, and clinical guidelines, to provide a comprehensive approach for treating late-life depression. It focused on dose optimization, treatment duration, and the selection of safer alternatives to anticholinergics and benzodiazepines (which can worsen cognitive function).

Results

The authors found that many older adults receive subtherapeutic doses of antidepressants, and they recommend first maximizing the dose (within the therapeutic range). They recommend switching to another class, such as a serotonin-norepinephrine reuptake inhibitor, bupropion, or mirtazapine, if the patient does not achieve remission after eight weeks at the highest tolerated dose. These alternatives may be beneficial for comorbidities like pain (duloxetine) or insomnia (mirtazapine). For patients who do not achieve remission after two trials from different classes, the authors advocate for aripiprazole augmentation, which has shown superior effectiveness compared to switching strategies. Although augmentation with lithium or switching to nortriptyline could be considered for highly resistant cases, these have lower remission rates, and the authors recommend considering ECT, transcranial magnetic stimulation, or esketamine.

The authors advocate for the regular use of scales like the PHQ-9 to help guide treatment adjustments and improve outcomes. After remission, they endorse continuing antidepressants for at least one year to prevent recurrence, with longer durations recommended for recurrent or severe cases.

The authors stress the importance of deprescribing inappropriate medications, particularly those with anticholinergic properties like diphenhydramine, tricyclic antidepressants, and paroxetine, which are tied to cognitive impairment and a higher risk of dementia. They also emphasize the need for careful tapering of benzodiazepines by reducing the dose by 10%–25% every 2 weeks to prevent withdrawal symptoms; and suggest a shift toward nonpharmacologic approaches for managing insomnia and anxiety, such as cognitive behavioral therapy, relaxation techniques, and improved sleep hygiene.

PRACTICE IMPLICATIONS

This review challenges two prevalent practices in geriatric depression care: 1) underdosing and 2) the tendency to "stick it out" with ineffective treatments. Just as for non-geriatric adults, don't settle for low doses, and don't shy away from switching.

Is Neuropsych Testing Better Than the MoCA for Diagnosing Mild Cognitive Impairment?

REVIEW OF: Weinstein AM et al, *Am J Geriatr Psychiatry* 2022;30(1):54–64

STUDY TYPE: Cross-sectional

WHEN WE WANT TO SCREEN an elderly patient for mild cognitive impairment (MCI), should we refer them to full-scale neuropsychological testing, or is it sufficient to use the Montreal Cognitive Assessment (MoCA), which we can do ourselves during an appointment? A recent study aimed to provide some guidance.

Researchers looked at baseline data from a study of older adults who had been enrolled in a large RCT. These patients had a history of major depressive disorder in remission, MCI, or both. An unblinded group of experts attempted to compare diagnoses using two approaches. The first used "gold-standard" neuropsychological testing; the second combined MoCA scores with DSM-5 criteria.

The study sample included 431 adults with a mean age of 71. The subjects were primarily White (78%), female (63%), and highly educated (74% had a 4-year college degree). Mean baseline score on the Montgomery-Åsberg Depression Rating Scale was 3.7, and the average MoCA score was 24.7.

Results

Although the researchers found moderate agreement between the diagnostic approaches ($p<0.0005$), there was discrepancy in 103 cases (23.8%). In 91 of those cases, neuropsych testing reported more severe cognitive impairment than the MoCA. Diagnostic discrepancies were more likely to occur in patients with a history of a major depressive episode or APOE4 carrier status, both of which are established risk factors for cognitive decline. Study limitations included a small sample size, lack of blinding, and lack of generalizability.

PRACTICE IMPLICATIONS

Although it remains the gold standard, most patients don't need neuropsychological testing to screen for MCI. The MoCA missed about a quarter of MCI cases, however, especially for patients with a history of depression or APOE4. You might also consider neuropsych testing when there is a family history of Alzheimer's disease, or when patients and family members report disparate degrees of impairment.

Linking Alzheimer's and Depression in Patients After 50

REVIEW OF: Wingo TS et al, *Alzheimers Dement* 2023;19:868–874

STUDY TYPE: Prospective cohort

LATE-LIFE DEPRESSION IS ASSOCIATED with a doubled risk of Alzheimer's disease (AD). There's also an association between early- and mid-life depression and dementia (Elser H et al, *JAMA Neurol* 2023;80(9):949–958). Depression and AD have some genetic correlations (Harerimana NV et al, *Biol Psychiatry* 2022;92(1):25–33). So, does late-life depression increase the risk of AD, or is depression an early sign of AD? This study addressed that question.

Researchers investigated 6,656 individuals who were 50 or older (median age 56) with normal cognition. All had European ancestry; 59% were women. The researchers assessed cognitive function and self-reported depression every 2 years for about 16 years. They also genotyped all subjects and calculated polygenic risk scores (PRS) for AD based on known genetic patterns.

Results

Investigators found that subjects with higher PRSs were more likely to report depression. This finding held up after adjusting for a genetic predisposition to depression, sex, age, and education. The association between AD and depression was not explained by the APOE4 allele, which is the strongest genetic risk factor for AD. This suggests that genetic variations contributing to AD risk may also contribute to late-life depression risk.

PRACTICE IMPLICATIONS

This study doesn't totally resolve the question of causality, or tell us if treating late-life depression influences AD risk. But it does tell us that for some in this relatively homogenous population, depression after 50 may indicate an underlying risk for AD. For your patients with depression after age 50, particularly those with a family history of AD, consider closer monitoring of cognition.

Lithium Therapy in Older Adults: Expert Recommendations

REVIEW OF: Christ J et al, *Pharmacopsychiatry* 2023;56(5):188–196

STUDY TYPE: Delphi survey

As the population ages, diagnoses of bipolar disorder (BD) in older adults are becoming more common, which means clinicians need to know how to use lithium safely in this group. This study provides some expert guidance.

The researchers conducted a Delphi survey, which is a structured method for reaching consensus among experts. They tapped into the expertise of 24 German specialists in geriatric medicine and lithium therapy.

Results

After 2 rounds of questionnaires, researchers reached a consensus on 21 key recommendations for starting, monitoring, and stopping lithium in older adults. These included:

1. **Starting lithium: Lithium is recommended for long-term maintenance of BD, as an add-on for treatment-resistant depression, and for suicide prevention in older adults.** There's no strict age cutoff for starting lithium. It can be used with medications like angiotensin-converting enzyme inhibitors, angiotensin receptor blockers, opioids, and diuretics—just make sure to monitor closely. Mild cognitive impairment and a history of falls are not contraindications for starting lithium. Check the patient's creatinine, estimated glomerular filtration rate (eGFR), blood count, electrolytes, thyroid-stimulating hormone (TSH), T3, T4, ECG, weight, blood pressure, and heart rate before prescribing lithium.

2. **Monitoring lithium: Regularly follow a patient's creatinine, eGFR, blood count, electrolytes, TSH, T3, T4, ECG, weight, and blood pressure.** Cystatin C, a protein that can give a more accurate picture of kidney function than creatinine alone, might be worth monitoring in certain cases. Depending on the situation, you might want to consider thyroid sonography, neurological examination, and psychological examination. The recommended serum lithium levels are 0.4–0.7 mmol/L for those aged 60–79, and 0.4–0.6 mmol/L for those 80 and up.

3. **Stopping lithium: If kidney function starts to decline, consult with a nephrologist.** When you do stop lithium, taper it slowly over three months to reduce the risk of a relapse. If you need an alternative, consider lamotrigine, valproate, or quetiapine.

Not everything was clear-cut—there wasn't consensus on using lithium during acute manic episodes, in patients with schizoaffective disorder, or alongside nonsteroidal anti-inflammatory drugs or digoxin. The experts also disagreed on how often to check lithium levels and eGFR when other medications are in the mix, and on specific contraindications like frailty, dementia, or being underweight.

PRACTICE IMPLICATIONS

These experts agree that lithium can be appropriate for mood stabilization and suicide prevention in older adults, but requires lower therapeutic doses and careful, individualized management. The gaps in their consensus show that there's still a lot we have to learn about safe and appropriate lithium use in this population.

Moderate Alcohol Use, Iron, and Cognitive Decline

REVIEW OF: Topiwala A et al, *PLOS Medicine* 2022;19(7):e1004039

STUDY TYPE: Prospective cohort

PRIOR RESEARCH HAS FOUND increased iron in the brains of heavy drinkers (Juhás M et al, *Neuroimage* 2017;148:115–122). However, for years it was thought that moderate alcohol intake was not harmful and could even enhance health. More recent research has called this into question. This study looked at the relationships between alcohol consumption, iron depositions, and cognitive function in a large, and relatively older, population.

The authors enrolled 20,729 people from the UK Biobank; their average age was 55, and 48.6% were women. Brain iron levels were measured using MRI, and an abdominal MRI checked for iron deposition in the liver. The authors controlled for the effects of occupation, education, hypertension, menopause, and other variables on iron levels. Alcohol use was assessed via questionnaires. Executive function, fluid intelligence, and reaction times were measured with the Trail-Making Test, puzzle tests, and a "snap" card game, respectively.

Results

Among the cohort, 95% self-identified as current drinkers, 2.4% had discontinued drinking, and only 2.7% were never drinkers. Their average alcohol consumption was about 10 standard drinks per week. Drinking more than four standard drinks weekly was associated with increased iron in the basal ganglia as well as poorer executive function, fluid intelligence, and reaction times. The authors found an alcohol-age interaction, suggesting that alcohol may magnify the effects of age on iron accumulation in the brain.

There were limitations to this study. For example, alcohol use was collected via self-report, and the models of iron in the brain were not robust enough to suggest it as a singular or primary cause of cognitive dysfunction. The authors also acknowledged that myelin may alter imaging markers, as iron and myelin share features on MRI.

PRACTICE IMPLICATIONS

This study supports an emerging consensus that less-than-heavy drinking is associated with health risks, in this case to executive function, fluid intelligence, and reaction times. Older adults can be encouraged to limit their alcohol intake to a maximum of four drinks or less weekly.

Smartphone Apps Benefit Memory and Quality of Life

REVIEW OF: Scullin MK et al, *J Am Geriatr Soc* 2022;70(2):459–469

STUDY TYPE: Uncontrolled trial

CAN SIMPLE SMARTPHONE APPS improve cognition in the elderly? While a Cochrane review found no high-quality studies showing a benefit for people with dementia (Van der Roest HG et al, *Cochrane Database Syst Rev* 2017;6(6):CD009627), the current study focused on people with mild cognitive impairment (MCI). Researchers were interested in evaluating prospective memory (PM, remembering things you've planned to do). A decline in PM is often an early marker of Alzheimer's disease and may lead to forgetting to take medications, pay bills, or attend appointments.

The authors recruited 52 older adults with an average age of 75, recently diagnosed with MCI or mild dementia, but independent in activities of daily living. Participants were mostly White, with an average of 14 years of education. The researchers randomly assigned them to use one of two smartphone apps for four weeks: a reminder app (Cortana) or a digital recording app (Voice Recorder for Android or Voice Memos for iPhone). Participants were taught how to use the apps to make their own reminders. They were then assigned PM tasks to complete on scheduled days and in certain locations. The researchers tested the participants' memory before and after the study using interviews and individualized memory tasks, like remembering to make a phone call on certain days. 90% of participants completed the study.

Results

Regardless of which app was used, participants performed better than expected on memory tasks. They completed PM tasks about 52% of the time, compared to a 20% expected task completion rate reported in prior similar studies. Two-thirds also had clinically significant improvements on both the Prospective-Retrospective Memory Questionnaire and a structured interview, which assessed performance on common daily activities requiring PM.

PRACTICE IMPLICATIONS

Although these apps were compensatory—"offloading" memory rather than enhancing it per se—they were associated with meaningful improvements in prospective memory performance and everyday functioning. Participants were educated, so their experience using a new technology may not be generalizable. Nevertheless, smartphone apps may be a relatively cheap intervention that helps cognitively impaired patients remember to tend to their bills, medications, and appointments.

Therapy in Dementia? Choose CBT

REVIEW OF: Orgeta V et al, *Cochrane Database Syst Rev* 2022;4(4):CD009125

STUDY TYPE: Systematic review and meta-analysis

WE OFTEN THINK about antidepressants before therapy when treating depression in patients with dementia. However, guidelines also recommend psychotherapy as a first-line treatment for mild and moderate depression. How do various psychological interventions compare in people with memory disorders?

This Cochrane review assessed the clinical efficacy of psychological interventions in reducing depression and anxiety in people living with mild cognitive impairment (MCI) or dementia. It included 29 trials between 1997 and 2020 with a total of 2,599 participants. Twenty-four trials were in people with diagnoses of dementia (most mild to moderate), and 5 were in people with MCI. There were 15 trials of cognitive behavioral therapies (4 cognitive behavioral therapy [CBT] trials, 8 behavioral activation trials, and 3 problem-solving therapy trials). CBT was modified to include cognitive strategies in early dementia and behavioral strategies in later stages. The authors also included 11 trials of supportive and counseling therapies, 3 trials of mindfulness-based cognitive therapy (MBCT), and 1 trial of interpersonal therapy (IPT). Control groups received either treatment as usual, attention-control education (information about educational groups available), or enhanced usual care (mood and cognitive problem assessment and referral to services). Treatment duration and intensity varied between studies.

Results

Cognitive behavioral therapies slightly outperformed treatment as usual or active control conditions in reducing depressive symptoms (standardized mean difference [SMD]=-0.23; 95% CI [-0.37, -0.10]) and improved remission rates (relative risk=1.84; 95% CI [1.18, 2.88]). Furthermore, CBT improved both quality of life (SMD=0.31; 95% CI [0.13, 0.50]) and activities of daily living (ADLs) (SMD=-0.25; 95% CI [-0.40, -0.09]). Most subjects had mild to moderate dementia, so these results are less certain for those with MCI or severe dementia. Supportive and counseling interventions had minimal effects on depressive or anxiety symptoms. Neither MBCT nor IPT had enough data to allow conclusions about efficacy.

PRACTICE IMPLICATIONS
In this review, CBT treated depression in people with dementia, but supportive interventions didn't really help. CBT improved depressive symptoms, quality of life, and ADLs, although the effects were relatively small.

MANAGING ADVERSE
EFFECTS

Antipsychotic Polypharmacy: Maybe Not So Risky After All?

REVIEW OF: Taipale H et al, *Am J Psychiatry* 2023;180(5):377–385

STUDY TYPE: Retrospective cohort

PATIENTS WITH SCHIZOPHRENIA often take multiple antipsychotics. The most recent APA guidelines for schizophrenia acknowledge that there is no evidence against polypharmacy. Observational studies have suggested that polypharmacy might reduce the risk of psychiatric hospitalization compared to monotherapy. How do polypharmacy and monotherapy compare in terms of more general risks?

A Finnish register-based study investigated the risk of nonpsychiatric and cardiovascular hospitalizations associated with antipsychotic polypharmacy and monotherapy. The study covered 61,889 patients with schizophrenia (average age 47) discharged from Finnish hospitals between 1972 and 2014. Antipsychotic dosages were calculated at each dispensing date and averaged to daily defined doses (DDDs).

Results

Between 1996 and 2017, during a median follow-up of 14.8 years, patients received monotherapy 46% of the time, polypharmacy 34% of the time, and no pharmacotherapy 20% of the time. About 46% were on high-dose ($\geq$1.6 DDD/day) monotherapy, and 53% were on high-dose polypharmacy. Comparing any polypharmacy against monotherapy showed no significant differences in nonpsychiatric (hazard ratio [HR]=0.99; 95% CI [0.97, 1.00]) or cardiovascular (HR=0.98; 95% CI [0.98, 1.05]) hospitalization risks. Interestingly, for individuals who had used both monotherapy and polypharmacy at medium to high dosages ($\geq$1.1 DDD/day), periods of polypharmacy were associated with up to a 13% reduction in risk of nonpsychiatric hospitalization compared to periods of similar monotherapy dosage use. Also, any polypharmacy was associated with a 6% reduced risk on the secondary outcome of psychiatric hospitalization (HR=0.94; 95% CI [0.92, 0.95]). The most commonly used two-drug combinations were olanzapine and quetiapine, risperidone and quetiapine, and clozapine and aripiprazole.

After omitting short exposures ($\leq$30 days) to polypharmacy, monotherapy, and nonuse, high-dose polypharmacy had a lower risk of nonpsychiatric hospitalization (HR=0.94; 95% CI [0.90, 0.97]) but did not confer a lower risk of cardiovascular hospitalization (HR=0.96; 95% CI [0.81, 1.13]) when compared to high-dose monotherapy.

The study's limitations included its reliance on prescriptions, which don't tell us if a medication was actually taken. Additionally, the use of long-acting injectables was higher in the high-dose polypharmacy group (39% vs 22% in high-dose monotherapy), potentially influencing the reduced psychiatric hospitalization rates. Also, the study did not state what other psychiatric drugs the patients were being prescribed, such as antidepressants or mood stabilizers.

PRACTICE IMPLICATIONS

This study found lower rates of nonpsychiatric and cardiovascular hospitalizations among patients receiving mid- to high-dose antipsychotic polypharmacy when compared to similar monotherapy dosages. The rate of psychiatric hospitalization was also slightly lower among the antipsychotic polypharmacy group. These results are only observational but do support the APA's most recent guideline condoning polypharmacy when necessary.

Lithium and Valproate Have Low (and Similar) Risks of Kidney Injury

REVIEW OF: Bosi A et al, *JAMA Netw Open* 2023;6(7):e2322056

STUDY TYPE: Retrospective cohort

STUDIES OF THE RELATIONSHIP between lithium and kidney injury have been mixed. A 2012 meta-analysis of lithium and kidney injury was inconclusive, and evidence generally has been conflicting. In an effort to clarify this issue, researchers once again looked at the risk of kidney injury in patients taking lithium—this time comparing it to those taking valproate.

This retrospective cohort study analyzed data from about 11,000 Stockholm residents who began lithium or valproate between 2007 and 2018 and had no prior history of kidney transplant or maintenance dialysis. Using data from Stockholm's Creatinine Measurements database, the study tracked these patients for up to 10 years, covering roughly 5,300 individuals per medication group, with a median follow-up duration of 4.5 years. Their primary outcome was progression of chronic kidney disease (CKD), incidence of acute kidney injury (AKI), changes in estimated glomerular filtration rate (eGFR), and the onset of new albuminuria. They measured both lithium and valproate levels, specifically looking at outcomes at lithium levels above and below 1.0 mmol/L.

Results

The study found no significant difference in CKD progression between patients starting on lithium or valproate, with about 3.5% of individuals in each group developing CKD. There were also no significant differences in non-CKD eGFR reduction, AKI risk, or albuminuria between the two groups. Surprisingly, the overall risk of AKI over the 10-year study period was actually 3.2% *lower* for lithium than valproate (95% CI [-5.6, -1.1]).

However, there was some concerning news about lithium. For the subgroup of lithium-treated patients with blood levels higher than 1.0 mmol/L, the risk of CKD was almost triple that of patients with lithium levels below 1.0. Even a lithium level of more than 0.8 mmol/L significantly increased the risk of AKI (hazard ratio=2.56, 95% CI [1.67, 3.92]).

PRACTICE IMPLICATIONS

Despite its reputation, in these Swedish patients, lithium did not increase the risk of acute or chronic kidney injury any more than valproate. The overall risk of kidney injury in both groups was low, and keeping lithium below 0.8 mmol/L (when clinically feasible) is safest renally. In this study, valproate was more likely to cause AKI than lithium. It does have a warning on the label for hypersensitivity reactions that may cause nephritis, and there are case reports of kidney injury in patients starting valproate (Anguissola G et al, *Pediatr Nephrol* 2023;38(6):1725–1731). It's not common practice to check kidney function when a patient starts valproate—but we should be aware of the rare possibility of AKI.

Strategies to Reduce Antipsychotic-Induced Hyperprolactinemia

REVIEW OF: Zhe L et al, *Transl Psychiatry* 2022;12(1):267

STUDY TYPE: Systemic review and meta-analysis

UP TO 70% OF PATIENTS on antipsychotic drugs face a common side effect: increased prolactin. When prolactin exceeds 20 ng/mL in nongravid patients, they can suffer from reduced libido, breast enlargement, and galactorrhea. Men may experience erectile dysfunction, women have an increased risk of breast cancers (see "Why Should We Care About Hyperprolactinemia With Antipsychotic Use?" immediately following this brief), and postmenopausal women may have decreased bone density.

A comprehensive review involving 31 RCTs, with around 2,000 participants in total, investigated ways that patients taking antipsychotics can lower elevated prolactin levels. The study explored switching antipsychotics, supplementing with vitamin B6 (200–1200 mg), using dopamine agonists such as bromocriptine and cabergoline, and even trying an herbal remedy (peony-glycyrrhiza decoction). Researchers also investigated the effectiveness of aripiprazole (which has been shown to lower prolactin in other studies) either added to or displacing other antipsychotics.

Results

Findings varied based on the initial prolactin levels of the patients. For those with very high levels (above 100 ng/mL), adding a low dose of aripiprazole (5 mg daily) was the only method that significantly cut prolactin levels. For levels between 50 and 100 ng/mL, switching to aripiprazole, adding aripiprazole as an adjunct (in doses of 5–10+ mg daily), or incorporating vitamin B6 significantly helped. When prolactin levels were below 50 ng/mL, none of the strategies made a significant difference. The study didn't report whether these reductions changed things clinically.

PRACTICE IMPLICATIONS

When prescribing antipsychotics, it's crucial to keep an eye on prolactin levels. For levels over 100 ng/mL, adding just 5 mg of aripiprazole can lower prolactin. For 50–100 ng/mL, adding or switching to aripiprazole, or adding high-dose B6, are all viable options. B6 is not without risks, and can cause neuropathy at doses beyond 1,000 mg daily. This study did not investigate reducing the offending antipsychotic, which may also be helpful.

Why Should We Care About Hyperprolactinemia With Antipsychotic Use?

REVIEW OF: Rahman T et al, *J Clin Psychopharmacol* 2022;42(1):7–16

STUDY TYPE: Retrospective cohort

WHEN WE SELECT antipsychotic medications for our patients, we typically consider adverse effects like extrapyramidal symptoms, weight gain, and diabetes. But what about the risk of cancer?

This large observational study (n=540,737) evaluated the risk of breast cancer in US women ages 18–64 prescribed antipsychotic drugs.

Results

When compared with women prescribed anticonvulsants and/or lithium, those prescribed antipsychotics had an overall 35% higher risk of breast cancer. For women taking highly prolactin-elevating antipsychotics (eg, haloperidol, risperidone, and paliperidone) and moderately prolactin-elevating antipsychotics (eg, lurasidone and olanzapine), the risks were 62% and 54%, respectively. Antipsychotics that produced mild or no prolactin elevation (eg, aripiprazole, clozapine, quetiapine, and ziprasidone) were not associated with an increased risk of breast cancer.

When analyses were stratified by age (18–50 vs 51–64), the study found the risk of breast cancer was much higher in younger women using moderately prolactin-elevating antipsychotics compared with older women on these medications (90% vs 40%).

Three contemporary or subsequent meta-analyses have also found an association between antipsychotics and breast cancer (Gao Z et al, *Front Oncol* 2022;12:993367; Leung JCN et al, *Epidemiol Psychiatr Sci* 2022;31:e61; de Moraes FCA et al, *BMC Cancer* 2024;24(1):712).

PRACTICE IMPLICATIONS

For female patients, particularly younger women and those with an elevated risk of breast cancer, we need to consider the risk of breast cancer that has been associated with antipsychotics that increase prolactin. When prolactin is elevated, reduce the dose, or switch to a lower-risk medication if possible (see table). For prolactin over 100 ng/mL, adding 5 mg of aripiprazole can be beneficial. For 50–100 ng/mL, high-dose vitamin B6 is an alternative (see "Strategies to Reduce Antipsychotic-Induced Hyperprolactinemia" immediately preceding this brief). Also, don't forget that prolactin can be elevated in male patients, too.

TABLE: Antipsychotic Risk of Hyperprolactinemia

Medications	Risk of Hyperprolactinemia
Chlorpromazine Fluphenazine Haloperidol Paliperidone Perphenazine Risperidone	Highest
Iloperidone Lurasidone Olanzapine	Lesser
Aripiprazole Asenapine Brexpiprazole Cariprazine Clozapine Quetiapine Ziprasidone	Lowest

MOOD DISORDERS

Adjuvant Clonidine for Mania

REVIEW OF: Ahmadphanah M et al, *J Psych Research* 2022;146:163–171

STUDY TYPE: RCT

IN THE 1980S, CLONIDINE SHOWED PROMISE in one open-label study and one small controlled trial of bipolar mania, but those findings weren't followed up until this study. Clonidine has serotonergic and dopaminergic effects as well as alpha agonism; it is FDA approved for ADHD, and is sedating. This study tested its effect on cognition during mania.

Researchers in Iran randomized 70 inpatients with acute mania to receive either adjuvant clonidine or placebo in addition to lithium (dosage range 900–1200 mg/day) over the course of 24 days. Clonidine was started at 0.2 mg/day and titrated toward a maximum of 0.6 mg/day (given as 2 divided doses in the evening). The primary outcome was improvement in mania as measured by the Young Mania Rating Scale (YMRS). Secondary outcomes were the Pittsburgh Sleep Quality Index (PSQI) and Mini-Mental State Examination (MMSE).

Results

Compared to placebo, those on adjuvant clonidine achieved significantly lower endpoint YMRS scores on day 24 (9.8 vs 13.6) and slept better (PSQI=4.5 vs 5.9), but no differences were found in cognition as measured by the MMSE. Clonidine's effect sizes were large for mania (0.9) and medium for sleep (0.6). Adverse effects were not reported, but typical tolerability issues with clonidine include sedation and hypotension; in this study, the drug was held if a participant's blood pressure dropped below 100/60 mmHg for 2 consecutive days. The majority (84%) of patients were male, which may limit the generalizability of the findings. Also, the MMSE is a relatively coarse instrument for measuring cognition, and it is quite possible that subtle differences in cognition were not elicited. Furthermore, the study failed to report 95% CIs for the outcomes and effect sizes of interest.

A later systematic review of 9 studies (n=222) of clonidine for bipolar mania found mixed results. While four non-randomized studies supported clonidine as an adjunct for mania, five placebo-controlled RCTs were inconsistent. The review also noted concerns about the potential precipitation of depression with clonidine in some bipolar patients, and higher dropout rates in clonidine groups. The authors concluded that there is low-grade evidence for clonidine as an adjuvant treatment for mania (Singal P et al, *Brain Sci* 2023;13(4):547).

PRACTICE IMPLICATIONS

Consider clonidine augmentation in mania when patients don't respond to or cannot tolerate conventional options, or when they have comorbidities like ADHD or hypertension, where clonidine has a more established role.

Brexpiprazole for Resistant Depression

REVIEW OF: Furukawa Y et al, *Psychiatry Clin Neurosci* 2022;76(9):416–422

STUDY TYPE: Meta-analysis

In 2015, BREXPIPRAZOLE WAS APPROVED as adjunctive therapy for major depression, with a wide dosing range from 0.5 to 3 mg. This meta-analysis synthesized existing studies to look for a sweet spot within that range.

The authors included six double-blind, randomized, placebo-controlled studies comparing brexpiprazole and placebo augmentation of antidepressants. Each lasted six weeks. Included patients (n=1671) had treatment-resistant depression, defined as inadequate response to 1–3 antidepressants, without other serious psychiatric comorbidities. The primary outcome was response (≥50% reduction in depression severity). The secondary outcomes were acceptability (dropouts for any reason) and tolerability (dropouts due to adverse effects). All of the studies were funded by the manufacturer of brexpiprazole.

Results

The benefits of brexpiprazole augmentation increased as the dose increased, until around 2 mg (odds ratio [OR]=1.52; 95% CI [1.12, 2.06]). After that, the dose-efficacy curve showed a slightly decreasing trend through the licensed dose of 3 mg (OR=1.40; 95% CI [0.95, 2.08]). The mean dose of brexpiprazole required to produce 50% of the maximum effect (ie, ED50) was 0.88 mg, and the mean dose to produce 95% of the maximum effect (ED95) was 1.79 mg. Dropouts for adverse events also peaked around 2 mg, and dropouts for any reason increased as the dose increased.

When compared to the corresponding rates for placebo, brexpiprazole augmentation at the ED95 is estimated to yield a response rate of 25% (95% CI [20%, 31%]), a dropout rate due to adverse events of 1% (95% CI [0%, 4%]), and an overall dropout rate of 14% (95% CI [10%, 20%]) at week 6.

PRACTICE IMPLICATIONS

When using brexpiprazole to augment antidepressants, aim for 1–2 mg. On average, doses above 2 mg lead to diminishing returns.

Cariprazine Augmentation for Major Depression

REVIEW OF: Sachs GS et al, *Am J Psychiatry* 2023;180(3):241–251

STUDY TYPE: RCT

WHEN A PATIENT with depression does not respond to monotherapy, adding a second-generation antipsychotic is an evidence-based augmentation strategy. Cariprazine, a D2, D3, and serotonin 5-HT1A partial agonist, was FDA approved in 2022 as an adjunct to antidepressants for the treatment of major depressive disorder (MDD). This industry-funded phase 3 trial was one of the key studies leading to that approval.

The 6-week, 7-country, double-blind study randomized 751 adults with MDD who had an inadequate response to monotherapy. One-third each received adjunctive cariprazine at 1.5 mg/day, 3 mg/day, or placebo. Participants had historically tried three or fewer antidepressants, but most had tried only one. No participants were using other psychiatric medications for affective symptoms during the trial period, and no one had attempted other options for treatment-resistant depression such as esketamine or ECT. Change in baseline was measured by the Montgomery-Åsberg Depression Rating Scale (MADRS).

Results

Those in the 1.5 mg group fared best, with a statistically significant mean 14.1-point drop in MADRS scores, compared to a mean 11.5-point drop among the placebo group. The 3 mg group had a mean 13.1-point drop that was not statistically significant when compared to placebo. Overall, response and remission did not differ between the three groups. Both cariprazine groups had double the rates of akathisia and nausea compared to placebo.

PRACTICE IMPLICATIONS

Cariprazine joins other second-generation antipsychotics as an option for MDD when one antidepressant doesn't do enough. However, the effect was modest and it's currently not available as a generic, so you may want to try other options first.

Dextromethorphan-Bupropion (Auvelity): A Novel Mechanism for Major Depression

REVIEW OF: Tabuteau H et al, *Am J Psychiatry* 2022;179(7):490–499

STUDY TYPE: RCT

Iᴛ's ꜰʀᴜsᴛʀᴀᴛɪɴɢ ꜰᴏʀ ʙᴏᴛʜ ᴘᴀᴛɪᴇɴᴛs ᴀɴᴅ ᴅᴏᴄᴛᴏʀs when depression treatments fall short, and unfortunately, this is a common problem. The seminal STAR*D trial, for example, found that two-thirds of depressed patients didn't achieve remission with first-line treatments, and those who responded only did so after eight weeks of treatment (Trivedi MH et al, *Am J Psychiatry* 2006;163:28–40). We need new treatment strategies that can improve, and accelerate, response rates.

Medications targeting the glutamatergic system appear promising. One such example is dextromethorphan, but it's metabolized too quickly to have a beneficial effect. Bupropion inhibits dextromethorphan's breakdown, so by pairing dextromethorphan with extended-release bupropion in one tablet, dextromethorphan's bioavailability and half-life increase sufficiently for it to have a therapeutic effect. (Bupropion's own antidepressant effects don't hurt either.)

This multisite, randomized, double-blind trial assessed a combination tablet of dextromethorphan-bupropion (AXS-05, now marketed as Auvelity) versus sustained-release bupropion in patients aged 18–65 (n=80) diagnosed with moderate to severe major depressive disorder (MDD). Participants received either dextromethorphan-bupropion (45 mg/105 mg tablet; n=43) or bupropion alone (105 mg tablet; n=37) once daily for the first 3 days, then twice daily for 6 weeks. Treatment responses were evaluated using weekly Montgomery-Åsberg Depression Rating Scale (MADRS) scores.

Results

Dextromethorphan-bupropion produced a significantly greater, and faster, reduction of depressive symptoms compared to bupropion alone. For example, at week 2, 47% of dextromethorphan-bupropion patients reached remission, versus 16% of bupropion patients. Response rates (defined as 50% decrease in MADRS score from baseline) were 61% with dextromethorphan-bupropion compared to 41% with bupropion at 6 weeks. The combo was well tolerated, with adverse events comparable to the bupropion group.

Limitations of the study included its short duration and the exclusion of patients with comorbid psychiatric or medical disorders. The study was also industry funded.

A 2023 review summarized eight clinical trials of dextromethorphan-bupropion for MDD, bipolar depression, and treatment-resistant depression. It concluded that the evidence was solid for MDD but mixed for the latter two conditions (Parincu Z and Iosifescu DV, *Expert Rev Neurother* 2023;23(3):205–212).

PRACTICE IMPLICATIONS

Dextromethorphan-bupropion is the only FDA-approved oral NMDA receptor antagonist for MDD. There's somewhat of a catch-22: Insurance often won't approve dextromethorphan-bupropion until patients have failed other treatments, yet the strongest evidence supports its use in MDD, not in treatment-resistant depression. Keep in mind that both dextromethorphan and bupropion have abuse potential and may not be a good choice for patients with substance use disorders.

Do Antidepressants Have a Role in Acute Bipolar Depression?

REVIEW OF: Hu Y et al, *Psychiatry Res* 2022;311:114468

STUDY TYPE: Systematic review and meta-analysis

ANTIDEPRESSANTS ARE THOUGHT to cause mania, mixed states, and rapid cycling in bipolar disorder, but evidence for this is mixed. A 2016 meta-analysis on this controversial subject found a small benefit for depressive symptoms (effect size 0.17), but no change to remission or response rates with short-term use (McGirr A et al, *Lancet Psychiatry* 2016;3(12):1138–1146). This paper updates that work with 13 additional trials.

The analysis included 19 RCTs that tested antidepressants as add-ons to mood stabilizers or antipsychotics in acute bipolar depression. The studies included 2,587 patients, 89% of whom had bipolar I; the mix included both inpatients and outpatients, and patients with substance use, psychotic symptoms, or rapid cycling. Most studies followed them for 6–10 weeks. Primary outcomes were a score of 1 or 2 on the Clinical Global Impression-Improvement subscale, or response as indicated by >50% improvement on rating scales. The researchers also assessed remission rates, time to remission, adverse reactions, mood switching, and dropouts.

Results

In the 16 studies that measured response, there was no significant difference between antidepressants and placebo. 59.2% of those taking antidepressants achieved response, compared with 51.2% on placebo (relative risk [RR]=1.10; 95% CI [0.98, 1.23]; moderate heterogeneity). Eleven studies measured remission and also found no significant difference; in those, 48.9% on antidepressants achieved remission, compared with 42.4% on placebo (RR=1.09; 95% CI [0.99, 1.20]; low heterogeneity). Mood switching occurred in about 5% of patients in both treatment and placebo arms. Neither arm proved better at preventing adverse reactions or dropouts.

Researchers then looked at whether antidepressants performed differently when paired with a traditional mood stabilizer (lithium, valproate, or carbamazepine) or an antipsychotic. The results for antipsychotics looked better, but two of the four antipsychotic trials were industry sponsored and larger than the others, weighing the statistical outcomes in their favor. The authors identified other quality concerns in about half the studies.

While this meta-analysis focused primarily on acute treatment, a 2023 RCT investigated maintenance with antidepressants. It included 177 patients with bipolar I disorder who had achieved remission from depression following treatment with escitalopram or bupropion extended-release alongside a mood stabilizer. Patients were randomized to either continue antidepressant treatment for 52 weeks or taper off the antidepressant (switching to placebo) at week 8. The study found that patients who continued antidepressant treatment had significantly fewer depressive relapses than those who discontinued

(17% vs 40%; hazard ratio=0.43; 95% CI [0.25, 0.75]), with a trend toward, but not a finding of, significantly more manic or hypomanic episodes (Yatham LN et al, *N Engl J Med* 2023;389(5):430–440).

PRACTICE IMPLICATIONS

In these studies, patients with bipolar I were not likely to flip into mania or a mixed mood episode when taking an antidepressant alongside a first-line bipolar medication, either acutely or in the long run. Escitalopram or bupropion helped fend off recurrences of depression, but also trended toward inducing manic or hypomanic periods. Stick with FDA-approved options for bipolar depression, and monitor patients carefully if they do take antidepressants.

Does Psychiatric Hospitalization Prevent Suicide Attempts?

REVIEW OF: Ross EL et al, *JAMA Psychiatry* 2024;81(2):135–143

STUDY TYPE: Retrospective cohort

WHEN ACUTELY SUICIDAL PATIENTS present to our emergency rooms and clinics, we may recommend (if not mandate) psychiatric hospitalizations. Does this actually reduce suicide attempts (SAs)?

Investigators looked at data from 196,610 visits to emergency departments and urgent care by veterans who had suicidal thoughts (SI) or SAs between 2010 and 2015. Patients were grouped based on their psychiatric diagnoses and the nature and timing of their suicidality. This was classified as SI only, an SA in the past two to seven days, or an SA in the past day. The primary outcome was whether patients had or SAs in the following year.

Results

About 71.5% of patients presenting with SI or SA were hospitalized; the rest were discharged. At 1 year, SA rates were nearly identical regardless of hospitalization—11.9% for hospitalized patients and 12.0% for those discharged.

However, a machine learning model revealed that the impact of hospitalization varied. It was associated with reduced SA risk in 28.1% of patients, but increased risk in 24.0%. Hospitalization offered no consistent benefit for patients with SI alone or an SA in the prior two to seven days—except in those with depression. In contrast, it significantly reduced SA risk (by 7%–9%) in patients who had SAs within the past 24 hours, regardless of diagnosis.

PRACTICE IMPLICATIONS

Assessment of suicidality requires a case-by-case consideration. Based on this study of veterans, hospitalization for suicidality should not be a reflex decision. Those most likely to benefit with respect to future SAs are patients with SAs the day before or, in the case of depression, within the last week. However, for those with SI or more remote SAs, the benefits of hospitalization are unclear and may even pose risks.

ECT and Risk of Suicide in Major Depression

REVIEW OF: Rönnqvist I et al, *JAMA Netw Open* 2021;4(7):e2116589l

STUDY TYPE: Retrospective cohort

W E KNOW ECT IS EFFECTIVE for patients with severe depression, but does it reduce the risk of suicide following an initial hospitalization?

A Swedish cohort study examined this question by evaluating rates of suicide in the year after ECT. Researchers enrolled patients who had their first hospitalization for the treatment of moderate depression, severe depression, or severe depression with psychosis (n=11,050). Half had received ECT; half had not. Non-ECT patients received standard treatments, which included pharmacotherapy with antidepressants and lithium. ECT was administered 3 times weekly on average, using unilateral electrode placement for 87% of patients.

Results

In the 12 months following hospital discharge, rates of suicide were significantly lower among patients who received ECT compared to patients who received other treatments (1.1% vs 1.6%; hazard ratio=0.72; 95% CI [0.52, 0.99]; p=0.04). ECT's greatest suicide-reducing benefit was for patients 65 or older (p=0.001) and patients with severe depression with psychosis (p=0.001). The rate of suicide was also significantly lower among ECT-treated patients ages 45–64 (p=0.05). In contrast, ECT was not associated with a suicide risk reduction among patients ages 18–44 (p=0.51) or patients with moderate depression (p=0.84).

Interestingly, all-cause mortality was also significantly lower in the ECT group within 3 (0.7% vs 2.9%) and 12 (1.7% vs 4.3%) months of discharge. The study did not include data about adverse events.

Two subsequent, and similar, studies have weighed in regarding ECT and suicidality. A retrospective cohort study of 67,000 inpatients with depression found that those who received ECT also had a reduced risk of subsequent suicide in the year after discharge (Kaster TS et al, *Lancet Psychiatry* 2022;9(6):435–446). But a study of hospitalized veterans receiving ECT found no difference within a year (Watts BV et al, *J Clin Psychiatry* 2022;83(3):21m13886).

PRACTICE IMPLICATIONS

It's well established that ECT protects against suicidality in the short term. This study, and at least one other, also show that compared to treatment as usual, ECT is associated with a lower risk of suicide in the year following a hospitalization for moderate or worse depression, at least among patients 45 and older.

L-Methylfolate Offers Modest Boost to Antidepressants

REVIEW OF: Maruf AA et al, *Pharmacopsychiatry* 2022;55(3):139–147

STUDY TYPE: Systematic review and meta-analysis

L-METHYLFOLATE IS A METABOLITE OF dietary folate and is FDA cleared as a "medical food" for the adjunctive treatment of depression. Approval of medical foods does not require the same level of rigor as medications. In this meta-analysis, researchers scrutinized the strength of the empirical evidence for L-methylfolate in depression.

A systematic literature search turned up only four RCTs of methylfolate as an adjunct to antidepressants. Most added methylfolate after failure of an antidepressant, usually a selective serotonin reuptake inhibitor (SSRI) or serotonin-norepinephrine reuptake inhibitor. Two trials enrolled patients with partial SSRI responses (n=223), while the others did not make this distinction (n=284). Three trials used 15 mg/day dosing; the fourth used 7.5 mg. Trial durations ranged from one to six months.

Results

Results showed a modest treatment effect on Hamilton Depression Rating Scale scores (standardized mean difference=0.38; 95% CI [-0.59, -0.17]; $p<.001$) and a similar effect on response rate (relative risk [RR]=1.26; 95% CI [1.07, 1.48]; $p=0.005$). Only the 15 mg dose was effective.

The analysis had several limitations. Of the included trials, two were industry sponsored, one did not clearly indicate the psychiatric medications that subjects were taking with adjunct L-methylfolate, and two did not report adverse events. The small number of published studies leaves open the possibility that unpublished studies with negative findings could attenuate the effect of L-methylfolate.

The authors did not include a large randomized controlled monotherapy trial (n=330) of Enlyte, presumably because this FDA-cleared product contains other vitamins in addition to 7.5 mg of L-methylfolate (folic acid; folinic acid; vitamins B1, B2, B3, B6, and B12; iron; magnesium; zinc; coenzyme Q10; and omega-3 fatty acids). Enlyte had a large effect size (0.88) as *monotherapy* in one clinical trial. The patients had moderate depression, some treatment-resistant and some not, but all were selected for polymorphisms of the MTHFR gene (C677T or A1298C), making it difficult to compare these results with the traditional L-methylfolate studies. In theory, those patients may be more likely to respond to L-methylfolate, but that theory has not been clinically tested, and we do not recommend routine genetic testing before starting L-methylfolate (Mech AW and Farah A, *J Clin Psychiatry* 2016;77(5):668–671).

PRACTICE IMPLICATIONS

L-methylfolate augments antidepressants with a modest effect size, similar to the effect of antipsychotic augmentation. It is a reasonable choice for patients who want a natural or well-tolerated option. The Enlyte trial suggests that this strategy is even more effective in patients with genetic impairments of folate metabolism (or that the extra ingredients in Enlyte provide a boost). L-methylfolate is available by prescription or over the counter (eg, Opti-Folate is available as 15 mg tablets for $8/month), while Enlyte is prescription only ($52/month if not covered by insurance, at www.enlyterx.com).

Magnetic Seizure Therapy: A Safer, Gentler Alternative to ECT?

REVIEW OF: Deng Z-D et al, *JAMA Psychiatry* 2024;81(3):240–249

STUDY TYPE: RCT

ECT is fast-acting and very effective for severe, treatment-resistant depression (TRD). However, despite refinements, it entails the risk of neurocognitive side effects. Magnetic seizure therapy (MST) also involves seizures. In this case, however, they're induced via transcranial magnetic stimulation and can be more focal than those from ECT. Several small studies have shown MST benefits depression. This large, multicenter, double-blind RCT pitted MST against ECT for TRD.

A total of 73 adults with severe TRD were enrolled. Of these, 38 were randomized to standard, ultra-brief pulse, right unilateral ECT, and 35 to MST. The treatment groups were equivalent in terms of anesthesia protocols, as well as in demographic characteristics such as age, sex, race/ethnicity, education level, and depression severity. The primary endpoint was change from baseline of total scores on the 24-item Hamilton Depression Rating Scale (HDRS-24). Response was defined as a reduction in HDRS-24 of at least 50%. Remission was defined as a reduction of at least 60% and a total score of 8 or less. Patients were followed for up to six months.

Results

No significant differences were seen between groups in rates of response (51% MST vs 43% ECT) or remission (45% MST vs 42% ECT). Sustained benefits across six months also were similar. The mean number of treatments to remission, however, was greater for MST (nine vs seven). This concerned the study authors because of the increased exposure to general anesthesia. They felt that further studies were warranted to optimize time to remission.

MST did have some benefits over ECT. Time to orientation was much more rapid, and autobiographical memory was sharper. Two cases of nausea/vomiting were reported following treatment with MST, whereas five serious adverse events occurred in the ECT group, including three cases of worsening depression, one case of increased blood pressure, and one case of prolonged ictal agitation.

PRACTICE IMPLICATIONS

This study suggests that MST is as effective as ECT. It requires an average of two more sessions than ECT, but it is also associated with less serious adverse events and fewer cognitive side effects.

Mitochondrial Modulators and Bipolar Depression

REVIEW OF: Liang L et al, *Transl Psychiatry* 2022;12(1):4

STUDY TYPE: Systematic review and meta-analysis

MITOCHONDRIAL DYSFUNCTION IS THOUGHT to play a role in bipolar disorder. Several controlled trials have examined whether nutritional supplements that support energy production in mitochondria can benefit bipolar depression. This meta-analysis gathered that evidence together for a big-picture review.

The authors identified 13 RCTs that investigated the antidepressive effects of specific mitochondria modulators in bipolar depression: N-acetylcysteine (NAC; 4 trials); omega-3 polyunsaturated fatty acids (3 trials); inositol (2 trials); and 1 trial each of coenzyme Q10 (CoQ10), creatine monohydrate, vitamin D, and acetyl-L-carnitine/alpha-lipoic acid combination. The primary outcome was the standardized mean difference (SMD) based on changes in the Montgomery-Åsberg Depression Rating Scale or Hamilton Depression Rating Scale. Using Cochrane guidelines, the authors determined that there was a low risk of publication bias among the studies. The total sample size was 605.

Results

Overall, the mitochondrial modulators significantly reduced depression severity compared to placebo, with a moderate effect size (SMD=0.48; 95% CI [0.14, 0.83]; p=0.007). However, only NAC and CoQ10 individually demonstrated significant reductions in depression severity. Since CoQ10 was only examined in a single study, the pooled effect size was mainly driven by NAC, with a wide CI around NAC's effect size (SMD=0.88; 95% CI [0.27, 1.48]; p=0.005). A wide CI means there is more uncertainty about the actual effect size of NAC. It could be as small as 0.27 or as large as 1.48, but the lower end of this range still indicates a positive effect on reducing depression severity. A possible explanation for this variance is that NAC can take a long time to work. The longer-term trials (greater than four months) tend to be positive, while the short-term studies tend to be negative.

PRACTICE IMPLICATIONS

We're not convinced that there is a class effect with mitochondrial agents, but NAC could be worth trying when other options fail in bipolar depression, especially if the patient prefers a non-medication approach. NAC is safe and well tolerated, with a recommended dosage of 2,000 mg daily.

Sublingual Dexmedetomidine (Igalmi) for Acute Agitation in Bipolar Disorder

REVIEW OF: Preskorn SH et al, *JAMA* 2022;327(8):727–736

STUDY TYPE: RCT

ACUTE AGITATION IS NOT UNCOMMON in mania, and sometimes it can't be managed without pharmacotherapy. Oral or even parenteral antipsychotics and benzodiazepines do not always work as quickly as we'd like, and they are poorly tolerated by some patients. It would be helpful to have additional options.

In April 2022, the FDA-approved dexmedetomidine sublingual film—an alpha-2 agonist used in intravenous form for procedural sedation and anesthesia—for the acute treatment of agitation in patients with bipolar disorder (BD) and schizophrenia. The sublingual formulation bypasses first-pass metabolism and is absorbed quickly.

This randomized, double-blind, placebo-controlled trial tested sublingual dexmedetomidine in patients with mild to moderate agitation associated with bipolar I and II disorder (n=380) across 15 clinical sites. The authors estimated baseline agitation by using the Positive and Negative Syndrome Scale-Excited Component (PEC) score, which includes 5 items (poor impulse control, tension, hostility, uncooperativeness, and excitement), each rated from 1 to 7. The PEC total score ranges from 5 (absence of agitation) to 35 (extremely severe agitation). Mean total PEC score at baseline was 18.

Patients were randomly assigned to receive either 120 or 180 μg of dexmedetomidine once, or placebo, self-administered under supervision of a staff member. A repeat dose of 60 or 90 μg could be given two hours after the first dose, at the investigators' discretion, if the change from baseline on the PEC scale was less than 40% and if there were no safety concerns. All patients also were continued on their current psychiatric medications.

Results

Dexmedetomidine was significantly more effective than placebo in reducing agitation (both dosages p<0.001 vs placebo). Its onset began within 20 minutes for both dosages. Response rates (defined as a decrease of 40% or more in PEC total score at 2 hours compared to baseline) were 91% (180 μg dosage), 77% (120 μg), and 46% (placebo). Unfortunately, the study did not compare dexmedetomidine with any standard agitation medications.

Dexmedetomidine produced no serious adverse events. The most common side effects were somnolence, dry mouth, hypotension, and dizziness. One patient in each of the dexmedetomidine groups reported suicidal ideation lasting one day.

PRACTICE IMPLICATIONS

Alpha-2 antagonism and sublingual film are both novel treatments for acute agitation—at least, mild to moderate agitation for which patients agree to take a medication. Sublingual dexmedetomidine was a fast-acting and well-tolerated option for these patients with BD. The sublingual film requires self-administration, but this may be less traumatic for patients than receiving parenteral medication.

Surprise Result for Adherence in Bipolar Disorder

REVIEW OF: Lintunen J et al, *J Affect Disord* 2023;333:403–408

STUDY TYPE: Retrospective cohort

PATIENTS OFTEN VOTE with their feet when it comes to medications. This study used data from a large national database to tell us which medications for bipolar disorder (BD) have the lowest and highest rates of nonadherence.

Researchers examined 33,131 patients from Finland's national health records database. All were under age 65 and diagnosed with BD between 1987 and 2018. Patients were included if they were prescribed mood stabilizers (lithium, valproic acid, or carbamazepine) or antipsychotics (27 agents were included) between 2015 and 2018. The primary outcome was nonadherence, defined as at least 1 medication, or 20% of total medications, going undispensed. The authors controlled for age, sex, time since BD diagnosis, and number of previous hospitalizations due to BD. They also controlled for other comorbidities, including mental health conditions like substance use disorders, anxiety, and personality disorders, as well as physical conditions like diabetes, cancer, and cardiovascular disease.

Results

Around one-third of patients were nonadherent to at least 20% of prescribed medications, and nearly twice that (59.1%) had at least 1 non-dispensed medication. Lithium fared best among the mood stabilizers, with 11.3% not dispensed compared to at least 14% for the other mood stabilizers. Clozapine beat all other antipsychotics, with 9% not dispensed compared to 13.5%–31.4% for others. (Rates of clozapine prescription are higher in Finland than in other countries.) Risks for nonadherence included psychiatric comorbidities, age under 25, diagnosis within past 3 years, benzodiazepines, and 4 or more hospitalizations.

Previous studies have found rates of nonadherence from 20% to 60% in BD, often relying on self-report. Using a more objective measure, this study arrived at the high end of that range (59.1%), but its result still may underestimate the problem as even filled prescriptions may not be taken.

PRACTICE IMPLICATIONS

This study is a reminder that patients stick with treatments that bring fuller recoveries, even when side effects might present an obstacle. Lithium is the gold standard in BD and has better preventative effects than other options. It is particularly effective in patients with classic, euphoric mania/hypomania, good functioning between episodes, and depressions that follow manias/hypomanias. Clozapine is used off-label for treatment-resistant mania and is used more commonly in Finland.

Top Augmentation Strategies for Treatment-Resistant Depression

REVIEW OF: Nuñez NA et al, *J Affect Disord* 2022;302:385–400

STUDY TYPE: Systematic review and meta-analysis

WHICH AUGMENTING AGENT is best when major depressive disorder (MDD) doesn't respond to monotherapy? This network meta-analysis aimed to find out.

The authors included 69 RCTs of adults 65 or younger with MDD not well managed by adequate doses of 1 or more antidepressants. All studies compared augmentation of an antidepressant, mainly selective serotonin reuptake inhibitors and tricyclic antidepressants, with an adjunct or placebo for up to 24 weeks. Researchers excluded studies focusing on specific patient populations, such as depression with psychotic features, bipolar depression, postpartum or prenatal depression, or MDD comorbid with a serious medical illness. The primary outcome was response rate, defined as a 50% decrease in Hamilton Depression Rating Scale or Montgomery-Åsberg Depression Rating Scale score. Secondary outcomes were remission and all-cause discontinuation rates. Results were expressed as relative risks (RRs), with values >1 indicating superiority.

Results

Significantly more effective than placebo were (in alphabetical order, as most differences were washed out by overlapping CIs): aripiprazole (RR=1.57; 95% CI [1.36, 1.82]), brexpiprazole (RR=1.56; 95% CI [1.15, 2.11]), cariprazine (RR=1.20; 95% CI [1.01, 1.42]), lisdexamfetamine (RR=1.18; 95% CI [1.03, 1.37]), lithium (RR=1.25; 95% CI [1.00, 1.56]), modafinil (RR=1.26; 95% CI [1.07, 1.48]), nortriptyline (RR=2.05; 95% CI [1.02, 4.11]), olanzapine (RR=1.23; 95% CI [1.00, 1.50]), quetiapine (RR=1.34; 95% CI [1.14, 1.56]), and T3 (RR=1.90; 95% CI [1.16, 3.11]). Those that did not reach significance were bupropion, buspirone, lamotrigine, methylphenidate, mirtazapine, pramipexole, risperidone, T4, and ziprasidone.

The antipsychotics had the most studies and the largest sample sizes (aripiprazole n=1147, brexpiprazole n=599, cariprazine n=963, lithium n=469, modafinil n=284, nortriptyline n=23, T3 n=114). Discontinuation rates were higher with cariprazine (RR=1.72; 95% CI [1.09, 2.73]), as well as with two that failed the significance test—ziprasidone (RR=20.12; 95% CI [1.17, 344.58]) and mirtazapine (RR=4.12; 95% CI [1.97, 8.63]).

Network meta-analyses compare medication and placebo arms from different studies, although measures of heterogeneity and inconsistency were insignificant in this analysis. The authors raised concerns about one result—lisdexamfetamine—as earlier meta-analyses were not as favorable toward this stimulant. Indeed, two large trials of lisdexamfetamine were negative, while the two smaller ones were positive. This study also focused on pharmacotherapy, which leaves out other effective augmentation strategies like lifestyle interventions (eg, exercise), psychotherapy, transcranial magnetic stimulation, ECT, and ketamine.

PRACTICE IMPLICATIONS

Antipsychotics were the best studied of the lot, but no particular medication was definitively most effective. The choice of an augmenting agent is best guided by balancing the benefits and risks of the medications with the severity of the patient's depression.

PSYCHOTHERAPY

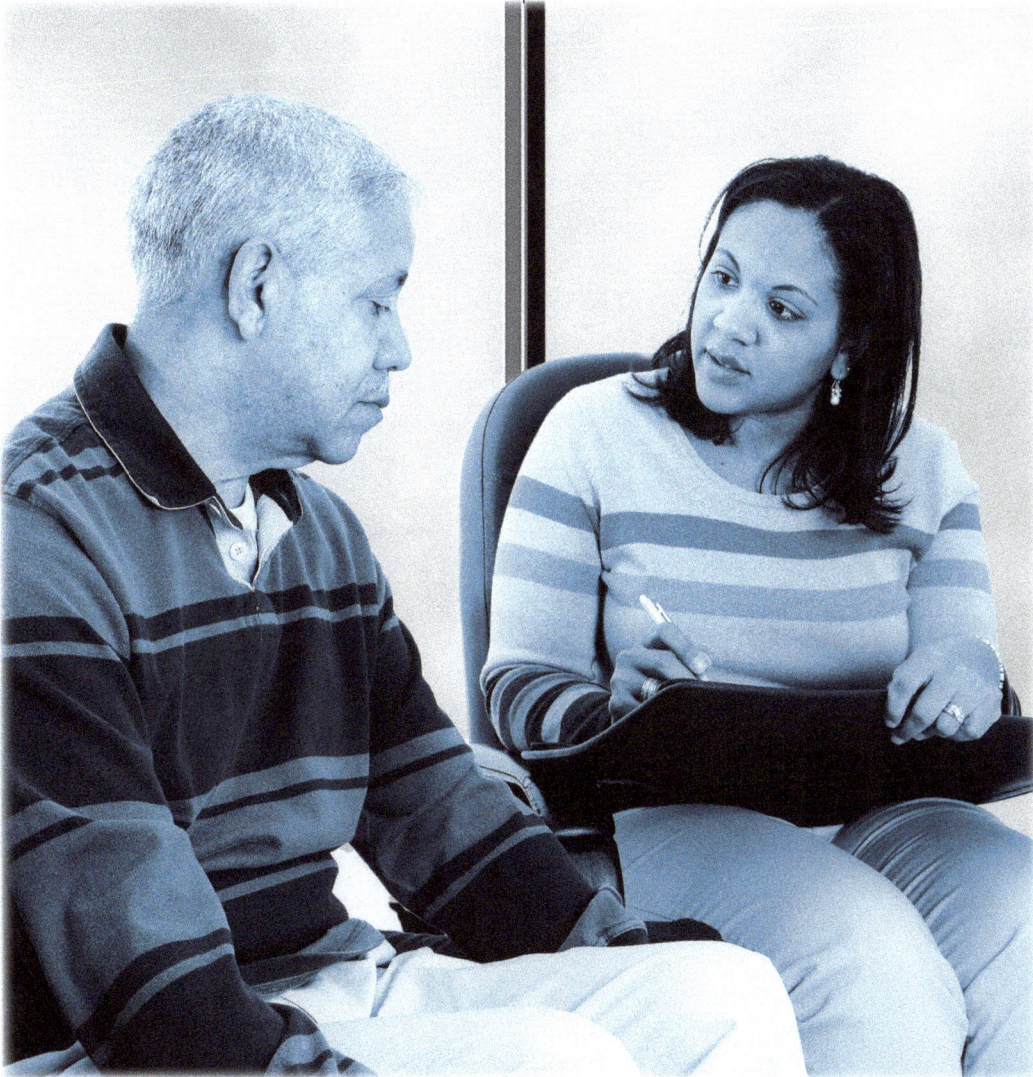

DBT for Borderline Personality Disorder: Is Half the "Dose" Effective?

REVIEW OF: McMain S et al, *Psychother Psychosom* 2022;91(6):382–397

STUDY TYPE: RCT

Most effective psychotherapies for borderline personality disorder (BPD) are long term, including dialectical behavior therapy (DBT). Many programs require patients to agree to a full year of DBT, which is a big commitment. Would a briefer course be effective? This study compared 6 versus 12 months of DBT for patients with BPD.

Researchers randomized 240 Canadian patients with BPD to receive either 6 months of DBT (DBT-6) or the standard treatment of 12 months (DBT-12). Patients were mostly women (79%) with a mean age of 28 and had to have a recent history of suicidal or nonsuicidal self-injurious episodes to be enrolled. DBT-6 and DBT-12 were identical except for length: Both involved the standard components, including weekly individual therapy, weekly group skills training, 24/7 phone coaching, and consultation meetings for therapists. The primary outcome was total frequency of self-harm episodes, assessed by the Suicide Attempt Self-Injury Interview at baseline and every 3 months for up to 24 months.

Results

The treatments did not differ significantly. At the beginning of the study, the mean number of self-harm events per patient in the preceding 3 months was 7.39. By the end of treatment for DBT-12 (which was 6 months post-treatment for DBT-6), mean self-harm events in the preceding 3 months had dropped to 0.30 (DBT-12) and 0.26 (DBT-6) with no significant difference between them. Mean self-harm events continued to decrease, to 0.22 (DBT-12) and 0.10 (DBT-6) at month 24. Dropout rates were similar for both groups. Secondary analyses showed DBT-12 patients having more improvement in interpersonal functioning and anger expression at month 24, but similar improvements in most other measures.

The study had several limitations. For one, there was no control arm, for ethical reasons. Additionally, concomitant treatments such as pharmacotherapy were not controlled for, and the primary outcome relied on self-report.

PRACTICE IMPLICATIONS

For patients with BPD and self-injury who hesitate to commit to a year of DBT, a shorter course may be just as effective in reducing self-harm. In areas where DBT resources are limited, clinicians can consider offering two 6-month courses in place of one 12-month course.

Enhancing PTSD Therapy With Aerobic Exercise: A New Approach

REVIEW OF: Bryant RA et al, *Lancet Psychiatry* 2023;10(1):21–29

STUDY TYPE: RCT

EXPOSURE THERAPY, A STAPLE PTSD TREATMENT, works through extinction learning. This involves repeated, safe exposure to trauma-related cues, which gradually reduces their perceived threat. However, up to half the patients exposed don't respond adequately. This led researchers to ask if exercise—which may enhance learning by raising brain-derived neurotrophic factor (BDNF)—could improve outcomes in exposure therapy.

The study recruited 130 participants (61% female; mean age about 39) with PTSD. About one-third were prescribed antidepressants. All were younger than 70, with a low suicide risk and no psychosis, substance dependence, brain injuries, or physical disorders that could interfere with exercise. About 50% of their traumas were classified as "assault," close to 30% as "road accidents," and the rest as other accidents. Participants were randomized to exposure therapy with either passive stretching or vigorous aerobic exercise. The exercise exposure was done 10 minutes after each of nine 90-minute weekly therapy sessions and was supervised by the therapists. They motivated participants to maintain their heart rates between 65% and 85% of their maximum (220 BPM minus age in years) as measured by chest or wrist monitors.

Results

After 6 months, both groups improved; however, the therapy and exercise group had a 12.1-point drop in symptoms (95% CI [2.4, 21.8]; p=0.023) on a blinded Clinician-Administered PTSD Scale (CAPS-2) compared to controls. This correlated with a moderate but meaningful effect size of 0.6 (95% CI [0.1, 1.1]). The exercise group also saw a more pronounced improvement in depression symptoms per the Beck Depression Inventory-II scale. There were no notable differences between groups in anxiety, alcohol use, or maladaptive appraisals.

Effects immediately after the nine-week interventions were not evident, and the study could not conclude that the observed improvements were due to BDNF. Nevertheless, these findings suggest that exercise may enhance long-term learning from exposure therapy. Some limitations of the study included most participants being White women, all being younger than 70, and the relatively strict inclusion criteria mentioned above.

PRACTICE IMPLICATIONS

It may not be practical for us to coach our patients ourselves (and it has interesting implications for the therapeutic frame), but whether they suffer from PTSD or anything else in the DSM, we can still encourage them to exercise. In this study, just 10 minutes of relatively vigorous aerobic exercise after each session of exposure therapy led to better outcomes for PTSD.

Is CBT Really All That Jazz for Depression?

REVIEW OF: Cuijpers P et al, *World Psychiatry* 2023;22(1):105–115

STUDY TYPE: Meta-analysis

GUIDELINES WIDELY RECOMMEND cognitive behavioral therapy (CBT) for depression because of its effectiveness versus control conditions. However, how does it perform compared to other treatments? This meta-analysis scrutinized studies where CBT went head-to-head against other psychotherapies, medications, or both.

It included 409 RCTs, encompassing 52,702 patients (27,000 in CBT and 25,702 in control groups) diagnosed with depression through diagnostic interviews or self-reports. When published, it was the largest meta-analysis ever conducted examining a specific psychotherapy for a particular mental disorder.

CBT was broadly defined as therapy centered on cognitive restructuring, regardless of its administration format or setting. Most studies enrolled adult participants (average age 40.1; 69% female), and were primarily conducted in the US, UK, and other European countries. Most studies were published after 2011, and individual-format CBT with more than 12 sessions was the most common format.

Results

CBT was significantly more effective than control conditions (usual care or waitlists) with an effect size of 0.79 (95% CI [0.70, 0.89]). This superiority persisted over 12 months of follow-up but tapered off at 13–24 months. Compared with other psychotherapies and accounting for biases, however, there was no significant difference. Similarly, CBT was not superior to pharmacotherapy or combined treatments after adjusting for publication bias. Medications with therapy did better than medications without.

PRACTICE IMPLICATIONS

This study confirms CBT's effectiveness for depression when compared to no treatment—however, there is no clear superiority over other psychotherapies, or over medication. If your CBT patients are struggling with homework assignments or other aspects of the treatment, it's likely that they will do just as well with other modalities, or with medications. This study also suggests that if you are prescribing antidepressants, your patients will do better if you include therapy of any kind.

Optimizing Sleep Timing for Night-Shift Workers

REVIEW OF: Cheng WJ et al, *Sleep* 2022;45(4):zsac034

STUDY TYPE: Randomized crossover trial

WORKING NIGHT SHIFTS is an occupational hazard for many. One in five shift workers have problems with sleep, attention, and wakefulness that meet criteria for shift-work disorder. Management of this disorder involves the appropriate timing of light and darkness, but no one had looked at the question of when sleep itself should occur. This study compared sleeping after work (that is, in the morning) to before it (at night) in those with shift-work disorder.

Sixty adults were randomized to a morning (9 AM–5 PM) or an evening (3 PM–11 PM) sleep schedule for the first week of the intervention, crossing over to the other schedule for the second week. Their average age was 30, 80% were women, and 66% worked as night-shift nurses. Adherence to the protocol was monitored with daily sleep and actigraphy data. No participant used prescribed sleep aids or stimulants during the study period.

Results

The evening sleep protocol resulted in a 30-minute longer duration of sleep (5.3 vs 4.8 hours) as well as improvements in sleep quality and daytime somnolence throughout the week. Attention also improved, but only on the day after the first shift. The evening protocol was particularly effective for subjects with an evening chronotype ("night owls").

PRACTICE IMPLICATIONS

Ask your patients who work the night shift whether they sleep before or after their shift. If a later sleep schedule is compatible with their social and family life, start a conversation about the potential benefits.

Psychological Benefits of Abstaining From Social Media

REVIEW OF: Lambert J et al, *Cyberpsychol Behav Soc Netw* 2022;25(5):287–293

STUDY TYPE: RCT

IT HAS LONG BEEN SPECULATED that spending too much time on social media might have negative effects on mental health. (Who wouldn't get depressed seeing how much fun everyone else seems to be having?) Although studies have consistently found a link between excessive social media use and depression, it can be hard to tell whether social media use exacerbates depression or is the result of it.

In this study, researchers recruited 154 volunteers from the community who agreed to be randomized to either continuing social media use as usual or abstaining from it for a week. The mean age of subjects was 29, and they spent on average just over an hour a day on social media. Subjects were not required to suffer from a mood or anxiety disorder to participate, although about one-third met criteria for moderate depression. The main outcomes of interest were reduction in depression and anxiety scores, as well as overall sense of well-being. Facebook, Instagram, Twitter, and TikTok were the four platforms the study focused on. Participants were provided with tips for abstaining, such as signing out of relevant social media sites, deleting apps, turning off notifications, turning off phones, and downloading app blockers. Screen time was monitored via apps.

Results

At least in the short term, abstaining from social media was indeed feasible. Subjects randomized to social media abstinence reduced their screen time use on average from 510 minutes to 21 minutes over the weeklong trial. This reduction was associated with improvements in depression, anxiety, and well-being scores, though the effect on depression was only apparent in those with at least mild depressive symptoms.

Limitations of the study included the recruitment of volunteers who were likely already motivated to abstain. Additionally, the study duration was only one week; the long-term ability of subjects to remain abstinent from social media and the associated potential psychiatric effects remain unknown.

A later meta-analysis found that social media abstinence was associated with a small to moderate reduction of depressive symptoms (standardized mean difference=-0.29; 95% CI [-0.51, -0.07]; p=0.01) in 3 RCTs. Abstinence did not yield improvements in life satisfaction, stress, or overall mental well-being, compared to continued use (Ramadhan RN et al, *Narra J* 2024;4(2):e786).

PRACTICE IMPLICATIONS

This study suggests that abstinence from social media use has beneficial effects on mental health. Researchers didn't describe what subjects were doing with their new-found free time.

There are many lifestyle changes, such as increasing exercise, getting adequate sleep, reducing substances, and practicing mindfulness, that we routinely recommend to our patients. It may be time to add reducing social media use to that list, particularly in our patients with depression.

SCHIZOPHRENIA

How Long Should We Wait Before Changing Antipsychotics in Schizophrenia?

REVIEW OF: Tang Y et al, *Curr Neuropharmacol* 2023;21(2):424–436

STUDY TYPE: Randomized, open-label trial

How long to wait before making changes after initiating an antipsychotic? The conventional wisdom is often four to six weeks, but emerging evidence indicates benefits may be observed as quickly as two weeks. This study tested whether early antipsychotic results predict subsequent efficacy.

In a multicenter, open-label trial, 3,010 Chinese patients with first-onset or relapsed schizophrenia were randomized to 7 antipsychotic drugs (risperidone, olanzapine, quetiapine, aripiprazole, ziprasidone, perphenazine, or haloperidol). They were monitored for six weeks using the Positive and Negative Syndrome Scale (PANSS) biweekly. Logistic regression and random forest models were used to determine if reductions in PANSS at week 2 and/or week 4 predicted a 50% or greater reduction in PANSS at week 6.

Results

Over 6 weeks, 56% achieved response. Two weeks was the earliest time point that predicted efficacy at 6 weeks. Results remained significant after adjustment for common sociodemographic (eg, age, education) and illness (eg, course, baseline severity) factors. No differences in predictive performance were observed between the antipsychotics. However, first-generation antipsychotics were better predictors of later efficacy than second-generation antipsychotics.

The strengths of the study included its large sample size. Limitations included confounding due to the many participants with relapsing versus first-onset illness.

PRACTICE IMPLICATIONS

See how patients with schizophrenia respond to a new antipsychotic at two weeks. If there's minimal improvement, consider switching or increasing, as the prospects for significant improvement later appear lower, especially for first-generation antipsychotics.

Immediate-Release Versus Extended-Release Quetiapine for Schizophrenia

REVIEW OF: Terao I et al, *J Psychopharmacology* 2023;37(10):953–959

STUDY TYPE: Systematic review and meta-analysis

QUETIAPINE IS WIDELY USED in varying doses and formulations to manage schizophrenia. Given its extensive dose range in both immediate-release (IR) and extended-release (ER) forms, determining which formulation is most appropriate to prescribe can be a challenge. This study offers a valuable comparison of IR and ER quetiapine across different dosages.

The study synthesized data from 6 double-blind RCTs involving 2,456 patients. This dose-response network meta-analysis allowed researchers to evaluate multiple doses within the same analysis, avoiding common issues found in traditional meta-analyses that mix different doses. The main outcomes were odds ratios of response rates on the Positive and Negative Syndrome Scale and/or the Brief Psychiatric Rating Scale.

Results

Both IR and ER quetiapine were more effective than placebo, which was not a surprise. However, there were significant differences across formulations and doses. IR quetiapine had a bell-shaped dose-response curve up to 500 mg, which peaked at a maximally effective dose of about 300 mg. ER quetiapine only had increasing efficacy above 300 mg, increasing linearly to a maximally efficacious dose around 550 mg, with efficacy decreasing minimally from 600 mg to 800 mg (the highest dose studied). Superimposing these dose-response curves showed IR significantly outperforming ER at doses between 100 and 300 mg, with ER better than IR in the 600–700 mg range. There was no significant difference in efficacy between IR and ER quetiapine at their doses of maximal response: approximately 300 mg and 600 mg, respectively.

The authors summarized their findings on quetiapine for schizophrenia with the following pair of recommendations:

- Begin with an IR formulation for doses up to 300 mg.
- If the clinical response at 300 mg of IR is inadequate, switch to ER to titrate further.

PRACTICE IMPLICATIONS

This meta-analysis provides us with some updated recommendations for prescribing quetiapine for schizophrenia. It suggests using IR quetiapine for doses up to 300 mg. For higher doses, ER is likely more effective, but may have diminishing returns above 600 mg.

Negative Symptoms of Schizophrenia: A Target for TMS?

REVIEW OF: Lorentzen R et al, *Schizophrenia (Heidelb)* 2022;8(1):35

STUDY TYPE: Systematic review and meta-analysis

THE NEGATIVE SYMPTOMS OF SCHIZOPHRENIA—low motivation, paucity of thought, impoverished speech, and poor self-care—are notoriously difficult to treat with medications. Transcranial magnetic stimulation (TMS) is FDA approved for similar presentations in major depression. A systematic review and meta-analysis synthesized the research on TMS for negative symptoms.

Included studies were double-blind, randomized, sham-controlled trials of TMS for adults with schizophrenia, schizoaffective disorder, or other psychotic disorders, with outcomes measured using the Positive and Negative Syndrome Scale or the Scale for Assessment of Negative Symptoms. Studies that initiated other treatments (eg, pharmacology) along with TMS were excluded. 57 studies (2,633 participants, 15 countries) met these criteria.

Results

The pooled analysis revealed the superiority of TMS over sham for reducing the negative symptoms of schizophrenia (Cohen's d=0.41; 95% CI [0.26, 0.56]; p<0.001; NNT=5). Subgroup analysis found that TMS aimed at the left dorsolateral prefrontal cortex (Cohen's d=0.55 vs standardized mean difference [SMD]=0.04; p=0.0002) and using a stimulation frequency greater than 1 Hz (Cohen's d=0.51 vs SMD=0.05; p=0.003) may be the most effective.

Limitations of the analysis included the substantial heterogeneity and risk for bias among the trials, including substantial differences in TMS modalities (ranging across unilateral, bilateral, rTMS, and deep TMS), total pulses (1,200–80,000), and different shams (coil repositioning vs sham coils vs undisclosed).

A 2024 JAMA Network dose-response meta-analysis of 14 sham-controlled studies (n=909) found a significant bell-shaped dose-response curve for negative symptoms in schizophrenia with left dorsolateral prefrontal cortex TMS. The curve peaked at 21,695 total pulses (95% CI [19,971, 23,531]). These findings indicate that there is most likely a range of total pulses in which maximal negative symptom response can be achieved, while total pulses substantially below and above this optimal range are associated with diminished efficacy. Like the Lorentzen et al paper, these findings are limited by considerable heterogeneity between included studies (Sabé M et al, *JAMA Netw Open* 2024;7(5):e2412616).

PRACTICE IMPLICATIONS
Despite a need for more consistent methodology in research, TMS shows promise for treating negative symptoms in schizophrenia spectrum disorders.

Pimavanserin and Negative Symptoms of Schizophrenia

REVIEW OF: Bugarski-Kirola D et al, *Lancet Psychiatry* 2022;9(1):46–58

STUDY TYPE: RCT

Could the negative symptoms of schizophrenia respond to a novel antipsychotic? Pimavanserin (Nuplazid) is approved for psychotic symptoms in Parkinson's disease. Rather than antagonizing dopamine D2 receptors, pimavanserin appears to work by blocking the 5-HT2A receptor, where it is both an antagonist and inverse agonist (as an antagonist, it blocks the receptor; as an inverse agonist, it binds to the receptor and induces the opposite effects as an agonist). An earlier trial found benefits for pimavanserin augmentation in major depression (Fava M et al, *J Clin Psychiatry* 2019;80(6):19m12928), and the current study was the first to test pimavanserin augmentation for negative symptoms of schizophrenia.

This was a randomized, double-blind, placebo-controlled trial of 403 stable, North American and European adult outpatients aged 18–55 (mean 38). All suffered from schizophrenia with predominant negative symptoms (specifically, they scored at least 20 on the negative symptom domain of the Positive and Negative Syndrome Scale). In addition to their current antipsychotic, patients were randomized to either pimavanserin (starting dose 20 mg daily; dose range 10–34 mg daily) or placebo for 26 weeks. The primary outcome was change in the Negative Symptom Assessment (NSA-16) from baseline to week 26. The secondary endpoint was the change in the Personal and Social Performance (PSP) scale, which assesses functioning in four areas: socially useful activities; personal and social relationships; self-care; and disturbing and aggressive behaviors. 86% of subjects completed the trial.

Results

After 26 weeks, the pimavanserin group showed a significantly greater reduction in negative symptoms versus placebo (-10.4 vs -8.5 points on the NSA-16), corresponding to a small effect size (0.21), but no change in PSP score (difference 0.0). In post-hoc analyses, the effect size for the NSA-16 was greater in patients who received the largest dose (34 mg) of pimavanserin (0.34), as well as in European males and patients with chronic negative symptoms (>5 years). Adverse effects were similar between study groups (35%–40%), most commonly headache and somnolence, and were typically mild. Pimavanserin was associated with a small but statistically significant increase in the QTc interval as compared to placebo (4.5 vs 0.0 ms). There were no clinically relevant effects of pimavanserin on vital signs, weight, glucose, or lipids in this six-month study.

However, a follow-up RCT by the same team failed to meet its primary endpoint. In that 26-week trial of 454 patients, pimavanserin at the higher 34 mg did not demonstrate statistically significant improvement over placebo on the NSA-16 total score (-11.8 vs -11.1; p=0.4825; effect size=0.07) (Bugarski-Kirola D et al, *Schizophr Bull* 2025; doi:10.1093/schbul/sbaf034).

PRACTICE IMPLICATIONS

These trials have conflicting results, and even the positive study didn't have a stellar effect size. We can only conclude that adjunctive pimavanserin is unlikely to significantly improve negative symptoms in schizophrenia. In the ongoing absence of an FDA-approved option, other medications with at least some evidence include cariprazine, olanzapine, clozapine, memantine, ondansetron, minocycline, xanomelium-trospium, and antidepressants (Govil P et al, *CNS Drugs* 2025;39(3):243–262).

WOMEN'S MENTAL HEALTH

Antipsychotics in Pregnancy and Risk of Neurodevelopmental Disorders

REVIEW OF: Straub L et al, *JAMA Intern Med* 2022;182(5):522–533

STUDY TYPE: Retrospective cohort

ANTIPSYCHOTICS CROSS THE PLACENTA, but data about the risk of congenital malformations in exposed children are generally reassuring (Huybrechts KF et al, *JAMA Psychiatry* 2016;73(9): 938–946). However, we know less about neurodevelopmental outcomes following prenatal antipsychotic exposures, particularly for second-generation antipsychotics.

A large retrospective US-based study examined the risk of neurodevelopmental disorders (eg, ADHD, autism spectrum disorder, learning disability, intellectual disability) in children following gestational exposure to antipsychotics. The study reviewed databases of publicly and privately insured women from 2000 to 2015. Children were considered exposed if their mothers filled antipsychotic prescriptions after 18 weeks of gestation. Why was the second half of pregnancy chosen? Synaptogenesis—the formation of nerve synapses—begins at this fetal stage.

Exposed (n=10,772) and unexposed (n=3,341,291) children were matched on potential confounders, including age, treatment indications, severity of maternal mental illnesses, adjunctive medications, maternal comorbidities, and socioeconomic status. Exposures were to first-generation and second-generation antipsychotics, with quetiapine being the most dispensed, followed by aripiprazole, risperidone, olanzapine, and haloperidol. Most women took only 1 antipsychotic during pregnancy, and their children were followed for up to 14 years.

Results

In unadjusted analyses, gestational exposure to antipsychotics was associated with a nearly twofold increase in the risk of neurodevelopmental disorders (hazard ratio [HR]=1.9). However, after adjusting for confounders, the increase was significant but much lower (HR=1.08; 95% CI [1.01, 1.17]). Among antipsychotics, only aripiprazole was linked with a small but statistically significant increased risk of neurodevelopmental disorders (HR=1.36; 95% CI [1.14, 1.63]).

A similar, more recent, cohort study from 5 Nordic countries examined 213,302 children (11,626 exposed to antipsychotics prenatally) with a median follow-up of 6.7 years. After the researchers adjusted for confounders, prenatal antipsychotic exposure generally—and exposure to any medication in particular, including aripiprazole—was not associated with an increased risk of neurodevelopmental disorders (HR=1.06; 95% CI [0.94, 1.20]). The study also found no increased risk for poor academic performance in mathematics or language arts (Bruno C et al, *eClinicalMedicine* 2024;70:102531).

PRACTICE IMPLICATIONS

These studies provide reassuring data about neurodevelopmental outcomes in children exposed to second-generation antipsychotics during the second half of pregnancy. The increased risk identified with prenatal aripiprazole was not replicated in the second study, but it's still probably better to avoid aripiprazole in pregnancy when possible.

Does Menopause Increase Psychosis Risk?

REVIEW OF: Sommer I et al, *Schizophr Bull* 2023;49(1):136–143

STUDY TYPE: Retrospective cohort

TIMES OF LOWER ESTROGEN, such as premenstrual and perimenopausal periods, can place females at risk of mood fluctuations. Are some women at higher risk for psychosis? This study examined whether females with schizophrenia spectrum disorders were more likely to relapse after age 45.

The authors identified nearly 62,000 inpatients with schizophrenia or schizoaffective disorder between 1972 and 2014 in Finnish nationwide registers. They stratified patients by sex and age, grouped ages into five-year periods, and compared doses for the four most common antipsychotic medications. Olanzapine was prescribed most for both sexes. For females, the second most commonly prescribed antipsychotic was risperidone, followed by quetiapine and clozapine.

Results

Up to age 45, females needed fewer hospitalizations for psychosis, and lower doses of antipsychotics, than males. After 45, females were significantly more likely to be hospitalized for psychosis than similarly aged males, and the gap widened as they got older. Older females were also more prone to relapse on standard doses of antipsychotics compared to women younger than 45, and to men of any age. Older females required higher antipsychotic doses for stabilization. Clozapine and olanzapine showed the largest drop in effectiveness as females aged past 45.

The study's strengths included its large sample size and the use of many subjects as their own controls within different age groups. The use of age to estimate menopause onset was a limitation, as was the lack of information about whether any females were also on estrogen supplementation.

The authors hypothesized that diminished estrogen levels heightened female patients' susceptibility to psychosis, and that menopause might influence the metabolism of antipsychotic medications. A 2024 study found that menopausal hormone therapy was associated with a 16% reduced risk of psychotic relapse in women with schizophrenia or schizoaffective disorder when used between ages 40–55, but not when started after age 56. This suggests that estrogen supplementation during the critical perimenopause window might be more effective than increasing antipsychotics (Brand BA et al, *Am J Psychiatry* 2024;181(10):893–900).

PRACTICE IMPLICATIONS

As your female patients with schizophrenia approach menopause, closely monitor them for signs of relapse, especially if they're on clozapine or olanzapine. During, and especially after, perimenopause, females with schizophrenia spectrum disorders appear particularly vulnerable to relapses. These two studies suggest they may benefit from higher antipsychotic doses or hormone replacement.

Perinatal Bipolar Mood Episodes: More Prevalent Than We Thought

REVIEW OF: Masters GA et al, *J Clin Psychiatry* 2022;83(5):21r14045

STUDY TYPE: Systematic review and meta-analysis

THE PERINATAL PERIOD increases the risk of bipolar exacerbations. This systematic review and meta-analysis of 22 observational studies involving perinatal women looked at the prevalence of bipolar spectrum disorders during pregnancy or postpartum.

Three studies included pregnant women, 9 studies looked at postpartum women, and 10 studies involved both. Subjects were classified as those without a known psychiatric history, or those with probable or known bipolar disorder (BD), assessed via diagnostic interviews or a diagnostic tool validated in perinatal populations.

Results

Among women with no prior psychiatric diagnosis, about 20% experienced a depressed, hypomanic, manic, or mixed mood episode during pregnancy or postpartum, and about 3% were diagnosed with BD. Among 2,814 women already diagnosed with BD, 55% had a bipolar mood episode during the perinatal period. Compared to those without a prior BD diagnosis, perinatal women who were identified with probable BD on a screening test were 6.5 times more likely to experience a depressive episode.

The main limitation here was that only about half of the studies incorporated a structured interview to establish a BD diagnosis. Additionally, since multiple studies reporting on women without previous psychiatric history looked only at depressive episodes, it is unclear how many in that population could go on to have lifelong bipolar illness.

PRACTICE IMPLICATIONS

For perinatal patients, be especially diligent when screening for BD and prescribing antidepressants. This is now a formal recommendation of the American College of Obstetricians and Gynecologists. In this meta-analysis, pregnancy amplified the risk of bipolar depression by nearly seven times.

Proposed Treatment Algorithm for Postpartum Psychosis

REVIEW OF: Jairaj C et al, *J Psychopharmacol* 2023;37(10):960–970

STUDY TYPE: Expert review

POSTPARTUM PSYCHOSIS (PPP) is a serious mental illness but not a diagnosis found in DSM-5-TR. This paper pulled together peer-reviewed English-language papers to review what is known about PPP and propose a treatment algorithm.

The prevalence of PPP is around 1–2 per 1,000 births. The strongest risk factors for PPP are bipolar disorder (BD) and a history of PPP. Having a first-degree relative with BD also increases the risk. Currently there are no validated PPP screeners available, but the Mood Disorder Questionnaire has been used.

Results

Women at risk should be offered preconception counseling, ideally from a specialist in perinatal mental health. For women with a history of PPP, researchers recommend prophylaxis with antipsychotics or lithium after delivery.

Mothers with PPP can develop symptoms very quickly—within hours—and most commonly between 3 and 10 days postpartum. Early symptoms include insomnia, anxiety, and mood fluctuations. Later, confusion, disorganized behavior, and mood changes are common.

Persecutory delusions, delusions of reference, visual hallucinations, and catatonia are also common. Visual hallucinations and delirium-like symptoms are more common with PPP than other psychoses. When diagnosing PPP, rule out postpartum blues and postpartum depression, which can present in similar ways. Infections, delirium, eclampsia, postpartum thyroiditis, autoimmune encephalitis, Sheehan syndrome (a postpartum pituitary disorder), and metabolic derangement should also be on your radar. Lastly, rule out medication- or substance-induced psychosis with a urine drug screen.

Other recommended labs include a complete blood count, metabolic profile, thyroid profile, urine drug screen, serum calcium levels, B12, folate, and thiamine. Rule out thyroiditis with antithyroid peroxidase antibodies and autoimmune encephalitis with anti-NMDA antibodies and imaging. One study also linked high C-reactive protein to PPP.

PPP is a psychiatric emergency and should trigger admission. Researchers recommend three medications: a second-generation antipsychotic, lithium, and short-term benzodiazepines to manage catatonia, agitation, or insomnia. For breastfeeding mothers, olanzapine and lorazepam are recommended due to their safety profile in lactation; consult a neonatologist about lithium. ECT is an option when a rapid response is needed or when medication doesn't work. The choice to breastfeed should be individualized, because it can affect medication selection—namely, lithium—and the sleep deprivation can exacerbate PPP.

No studies of antidepressants or psychosocial interventions in PPP were identified. Individual and couples therapy is suggested for addressing emotions and relationship challenges precipitated by an episode of PPP.

PRACTICE IMPLICATIONS
PPP is a life-threatening condition that requires early screening and intervention, along with inpatient treatment. This algorithm puts lithium, antipsychotics, and benzodiazepines first line, with ECT reserved for refractory cases or a rapid response.

9798989326471